Classical Summability Theory

P.N. Natarajan

Classical Summability Theory

P.N. Natarajan
Formerly of the Department of Mathematics
Ramakrishna Mission Vivekananda College
Chennai, Tamil Nadu
India

ISBN 978-981-13-5077-1 ISBN 978-981-10-4205-8 (eBook)
DOI 10.1007/978-981-10-4205-8

Softcover re-print of the Hardcover 1st edition 2017

Printed on acid-free paper

This Springer imprint is published by Springer Nature
The registered company is Springer Nature Singapore Pte Ltd.
The registered company address is: 152 Beach Road, #21-01/04 Gateway East, Singapore 189721, Singapore

Dedicated to my wife Vijayalakshmi and my children Sowmya and Balasubramanian

Preface

The study of convergence of infinite series is a very old art. In ancient times, people were interested in orthodox examination for convergence of infinite series. Divergent series, i.e., infinite series which do not converge, was of no interest to them until the advent of L. Euler (1707–1783), who took up a serious study of divergent series. He was later followed by a galaxy of very great mathematicians.

Study of divergent series is the foundation of summability theory. Summability theory has many utilities in analysis and applied mathematics. An engineer or physicist, who works on Fourier series, Fourier transforms or analytic continuation, can find summability theory very useful for his/her research.

In the present book, some of the contributions of the author to classical summability theory are highlighted, thereby supplementing, the material already available in standard texts on summability theory.

There are six chapters in all. The salient features of each chapter are listed below. In Chap. 1, after a very brief introduction, we recall well-known definitions and concepts. We state and prove Silverman–Toeplitz theorem, Schur's theorem and then deduce Steinhaus theorem. We introduce a sequence space Λ_r, $r \geq 1$ being a fixed integer and make a detailed study of the space Λ_r, especially from the point of view of sequences of zeros and ones. We prove a Steinhaus type result involving the space Λ_r, which improves Steinhaus theorem. We prove some more Steinhaus type theorems too.

Chapter 2 deals with the core of a sequence. We present an improvement of Sherbakhoff's result, which leads to a short and very elegant proof of Knopp's core theorem. We also present some nice properties of the class (ℓ, ℓ) of infinite matrices.

Chapter 3 is devoted to a detailed study of some special methods of summability, viz., the Abel method, the Weighted mean method, the Euler method and the (M, λ_n) or Natarajan method. We bring out the connection between the Abel method and the Natarajan method. Some product theorems involving certain summability methods are also proved.

In Chap. 4, some nicer properties of the (M, λ_n) method are established. Further, we prove a few results on the Cauchy multiplication of certain summable series.

In Chap. 5, a new definition of convergence of a double sequence and a double series is introduced. In the context of this new definition, Silverman–Toeplitz theorem for 4-dimensional infinite matrices is proved. We also prove Schur's and Steinhaus theorems for 4-dimensional infinite matrices.

Finally in Chap. 6, we introduce the Nörlund, the Weighted mean and the $(M, \lambda_{m,n})$ or Natarajan methods for double sequences and double series and study some of their properties.

I thank my mentor Prof. M.S. Rangachari for initiating me to the topic of summability theory—both classical and ultrametric. I thank Mr. E. Boopal for typing the manuscript.

Chennai, India P.N. Natarajan

Contents

About the Author

P.N. Natarajan formerly with the Department of Mathematics, Ramakrishna Mission Vivekananda College, Chennai, India, has been an independent researcher and mathematician since his retirement in 2004. He did his Ph.D. at the University of Madras, under Prof. M.S. Rangachari, former director and head of the Ramanujan Institute for Advanced Study in Mathematics, University of Madras. An active researcher, Prof. Natarajan has published over 100 research papers in several international journals like the *Proceedings of the American Mathematical Society*, *Bulletin of the London Mathematical Society*, *Indagationes Mathematicae*, *Annales Mathematiques Blaise Pascal* and *Commentationes Mathematicae* (Prace Matematyczne). His research interests include summability theory and functional analysis (both classical and ultrametric). Professor Natarajan was honored with the Dr. Radhakrishnan Award for the Best Teacher in Mathematics for the year 1990–1991 by the Government of Tamil Nadu. In addition to being invited to visit several renowned institutes in Canada, France, Holland and Greece, Prof. Natarajan has participated in several international conferences and chaired sessions. He has authored two books, *An Introduction to Ultrametric Summability Theory* and its second edition, both published with Springer in 2013 and 2015, respectively.

Chapter 1
General Summability Theory and Steinhaus Type Theorems

The present chapter is devoted to a study of some basic concepts in summability theory and Steinhaus type theorems. This chapter is divided into 5 sections. In the first section, we introduce some basic definitions and concepts. In the second section, we prove the Silverman–Toeplitz theorem, Schur's theorem, and Steinhaus theorem. In the third section, we introduce the sequence space Λ_r, $r \geq 1$ being a fixed integer and prove a Steinhaus type result improving Steinhaus theorem. The fourth section is devoted to a study of the sequence spaces Λ_r. In the final section, we prove more Steinhaus type theorems.

1.1 Basic Definitions and Concepts

The study of convergence of infinite series is an ancient art. In ancient times, people were more concerned with orthodox examinations of convergence of infinite series. Series, which did not converge, was of no interest to them until the advent of L. Euler (1707–83), who took up a series study of "divergent series," i.e., series which did not converge. Euler was followed by a galaxy of great mathematicians C.F. Gauss (1777–1855), A.L. Cauchy (1789–1857), and N.H. Abel (1802–29). The interest in the study of divergent series temporarily declined in the second half of the nineteenth century. It was rekindled at a later date by E. Cesàro, who introduced the idea of $(C, 1)$ convergence in 1890. Since then, many other mathematicians have been contributing to the study of divergent series. Divergent series have been the motivating factor for the introduction of summability theory.

Summability theory has many uses in analysis and applied mathematics. An engineer or a physicist, who works with Fourier series, Fourier transforms, or analytic continuation can find summability theory very useful for his/her research.

Consider the sequence

$$\{s_n\} = \{1, 0, 1, 0, \dots\},$$

P.N. Natarajan, *Classical Summability Theory*, DOI 10.1007/978-981-10-4205-8_1

which is known to diverge. However, let

$$t_n = \frac{s_0 + s_1 + \cdots + s_n}{n+1}, \quad n = 0, 1, 2, \ldots,$$

$$\begin{aligned} i.e., \ t_n &= \frac{k+1}{2k+1}, \ \text{if } n = 2k; \\ &= \frac{k+1}{2k+2}, \ \text{if } n = 2k+1, \end{aligned}$$

proving that

$$t_n \to \frac{1}{2}, n \to \infty.$$

In this case, we say that the sequence $\{s_n\}$ converges to $\frac{1}{2}$ in the sense of Cesàro or $\{s_n\}$ is $(C, 1)$ summable to $\frac{1}{2}$. Similarly, consider the infinite series

$$\sum_{n=0}^{\infty} a_n = 1 - 1 + 1 - 1 + \cdots.$$

The associated sequence $\{s_n\}$ of partial sums is $\{1, 0, 1, 0, \ldots\}$, which is $(C, 1)$ summable to $\frac{1}{2}$. In this case, we say that the series $1 - 1 + 1 - 1 + \cdots$ is $(C, 1)$ summable to $\frac{1}{2}$.

We now recall the following definitions and concepts.

Definition 1.1 Given an infinite matrix $A = (a_{nk}), n, k = 0, 1, 2, \ldots$ and a sequence $x = \{x_k\}$, $k = 0, 1, 2, \ldots$, by the A-transform of $x = \{x_k\}$, we mean the sequence $Ax = \{(Ax)_n\}$,

$$(Ax)_n = \sum_{k=0}^{\infty} a_{nk} x_k, \quad n = 0, 1, 2, \ldots,$$

where we suppose that the series on the right converge. If $\lim_{n\to\infty} (Ax)_n = s$, we say that the sequence $x = \{x_k\}$ is A-summable or summable A to s.

Given an infinite series $\sum_{k=0}^{\infty} x_k$, define

$$s_n = \sum_{k=0}^{n} x_k, \quad n = 0, 1, 2, \ldots.$$

If $\{s_n\}$ is A-summable to s, we say that $\sum_{k=0}^{\infty} x_k$ is A-summable to s.

Definition 1.2 Let X, Y be sequence spaces. The infinite matrix $A = (a_{nk})$, $n, k = 0, 1, 2, \ldots$ is said to transform X to Y, written as $A \in (X, Y)$ if whenever the sequence $x = \{x_k\} \in X$, $(Ax)_n$ is defined, $n = 0, 1, 2, \ldots$ and the sequence $\{(Ax)_n\} \in Y$.

Definition 1.3 Let c be the Banach space consisting of all convergent sequences with respect to the norm $\|x\| = \sup_{k\geq 0} |x_k|$, $x = \{x_k\} \in c$. If $A = (a_{nk}) \in (c, c)$, A is said to be convergence preserving or conservative. If, in addition, $\lim_{n\to\infty} (Ax)_n = \lim_{k\to\infty} x_k$, $x = \{x_k\} \in c$, A is called a regular matrix or a Toeplitz matrix. If A is regular, we write

$$A \in (c, c; P),$$

the letter P denoting "Preservation of limit."

1.2 The Silverman–Toeplitz Theorem, Schur's Theorem, and Steinhaus Theorem

We now prove a landmark theorem in summability theory, due to Silverman–Toeplitz, which characterizes a regular matrix in terms of the entries of the matrix (see [1–3]).

Theorem 1.1 (Silverman–Toeplitz) *$A = (a_{nk})$ is regular, i.e., $A \in (c, c; P)$ if and only if*

$$\sup_{n\geq 0} \sum_{k=0}^{\infty} |a_{nk}| < \infty; \tag{1.1}$$

$$\lim_{n\to\infty} a_{nk} = 0, \; k = 0, 1, 2, \ldots; \tag{1.2}$$

and

$$\lim_{n\to\infty} \sum_{k=0}^{\infty} a_{nk} = 1. \tag{1.3}$$

Proof Proof of the sufficiency part. Let (1.1), (1.2), and (1.3) hold. Let $x = \{x_k\} \in c$ with $\lim_{k\to\infty} x_k = s$. Since $\{x_k\}$ converges, it is bounded and so $|x_k| \leq M$, $k = 0, 1, 2, \ldots, M > 0$. Now,

$$\sum_{k=0}^{\infty} |a_{nk}x_k| \leq M \sum_{k=0}^{\infty} |a_{nk}| < \infty, \text{ in view of (1.1)}$$

and so

$$(Ax)_n = \sum_{k=0}^{\infty} a_{nk}x_k \text{ is defined, } n = 0, 1, 2, \ldots.$$

Now,

$$(Ax)_n = \sum_{k=0}^{\infty} a_{nk}(x_k - s) + s \sum_{k=0}^{\infty} a_{nk}, \; n = 0, 1, 2, \ldots. \tag{1.4}$$

Since $\lim_{k \to \infty} x_k = s$, given $\epsilon > 0$, there exists $n \in \mathbb{N}$, where $\mathbb{N}$ denotes the set of all positive integers, such that

$$|x_k - s| < \frac{\epsilon}{2L}, \; k > N, \tag{1.5}$$

where $L > 0$ is such that

$$|x_n - s| \leq L, \; \sum_{k=0}^{\infty} |a_{nk}| \leq L, \; n = 0, 1, 2, \ldots. \tag{1.6}$$

Thus,

$$\sum_{k=0}^{\infty} a_{nk}(x_k - s) = \sum_{k=0}^{N} a_{nk}(x_k - s) + \sum_{k=N+1}^{\infty} a_{nk}(x_k - s),$$

$$\left| \sum_{k=0}^{\infty} a_{nk}(x_k - s) \right| \leq \sum_{k=0}^{N} |a_{nk}||x_k - s| + \sum_{k=N+1}^{\infty} |a_{nk}||x_k - s|.$$

$$\begin{aligned}
\sum_{k=N+1}^{\infty} |a_{nk}||x_k - s| &\leq \frac{\epsilon}{2L} \sum_{k=0}^{\infty} |a_{nk}|, \text{ using (1.5)} \\
&\leq \frac{\epsilon}{2L} L, \text{ using (1.6)} \\
&= \frac{\epsilon}{2};
\end{aligned}$$

Using (1.2), we have

$$|a_{nk}| < \frac{\epsilon}{2L(N+1)}, \; k = 0, 1, \ldots, N,$$

so that,

$$\begin{aligned}
\sum_{k=0}^{N} |a_{nk}||x_k - s| &< L(N+1) \frac{\epsilon}{2L(N+1)} \\
&= \frac{\epsilon}{2}.
\end{aligned}$$

Consequently,

$$\left|\sum_{k=0}^{\infty} a_{nk}(x_k - s)\right| < \frac{\epsilon}{2} + \frac{\epsilon}{2}$$
$$= \epsilon, \text{ for every } \epsilon > 0.$$

Thus,

$$\lim_{n\to\infty} \sum_{k=0}^{\infty} a_{nk}(x_k - s) = 0. \tag{1.7}$$

Taking limit as $n \to \infty$ in (1.4), we have

$$\lim_{n\to\infty} (Ax)_n = s, \text{ using (1.3) and (1.7)}$$

and so A is regular, completing the proof of the sufficiency part.

Proof of the necessity part. Let A be regular. For every fixed $k = 0, 1, 2, \ldots,$ consider the sequence $x = \{x_n\}$, where

$$x_n = \begin{cases} 1, & n = k; \\ 0, & \text{otherwise.} \end{cases}$$

For this sequence x, $(Ax)_n = a_{nk}$. Since $\lim_{n\to\infty} x_n = 0$ and A is regular, it follows that

$$\lim_{n\to\infty} a_{nk} = 0, \; k = 0, 1, 2, \ldots$$

so that (1.2) holds. Again consider the sequence $x = \{x_n\}$, where $x_n = 1$, $n = 0, 1, 2, \ldots$. Note that $\lim_{n\to\infty} x_n = 1$. For this sequence x, $(Ax)_n = \sum_{k=0}^{\infty} a_{nk}$. Since $\lim_{n\to\infty} x_n = 1$ and A is regular, we have

$$\lim_{n\to\infty} \sum_{k=0}^{\infty} a_{nk} = 1,$$

so that (1.3) also holds. It remains to prove (1.1). First, we prove that $\sum_{k=0}^{\infty} |a_{nk}|$ converges, $n = 0, 1, 2, \ldots$. Suppose not. Then, there exists $N \in \mathbb{N}$ such that

$$\sum_{k=0}^{\infty} |a_{Nk}| \text{ diverges.}$$

In fact, $\sum_{k=0}^{\infty} |a_{Nk}|$ diverges to ∞. So, we can find a strictly increasing sequence $k(j)$ of positive integers such that

$$\sum_{k=k(j-1)}^{k(j)-1} |a_{Nk}| > 1, \ j = 1, 2, \dots . \tag{1.8}$$

Define the sequence $x = \{x_k\}$, where

$$x_k = \begin{cases} \frac{|a_{Nk}|}{j a_{Nk}}, & \text{if } a_{Nk} \neq 0 \text{ and } k(j-1) \leq k < k(j), \ j = 1, 2, \dots; \\ 0, & \text{if } k = 0 \text{ or } a_{Nk} = 0. \end{cases}$$

Note that $\lim_{k \to \infty} x_k = 0$ and $\sum_{k=0}^{\infty} a_{nk} x_k$ converges, $n = 0, 1, 2, \dots$. In particular, $\sum_{k=0}^{\infty} a_{Nk} x_k$ converges. However,

$$\begin{aligned} \sum_{k=0}^{\infty} a_{nk} x_k &= \sum_{j=1}^{\infty} \sum_{k=k(j-1)}^{k(j)-1} \frac{|a_{Nk}|}{j} \\ &= \sum_{j=1}^{\infty} \frac{1}{j} \sum_{k=k(j-1)}^{k(j)-1} |a_{Nk}| \\ &> \sum_{j=1}^{\infty} \frac{1}{j}. \end{aligned}$$

This leads to a contradiction since $\sum_{j=1}^{\infty} \frac{1}{j}$ diverges. Thus,

$$\sum_{k=0}^{\infty} |a_{nk}| \text{ converges, } \ n = 0, 1, 2, \dots .$$

To prove that (1.1) holds, we assume that

$$\sup_{n \geq 0} \sum_{k=0}^{\infty} |a_{nk}| = \infty$$

and arrive at a contradiction.

We construct two strictly increasing sequences $\{m(j)\}$ and $\{n(j)\}$ of positive integers in the following manner.

Let $m(0) = 0$. Since $\sum_{k=0}^{\infty} |a_{m(0),k}| < \infty$, choose $n(0)$ such that $\sum_{k=n(0)+1}^{\infty} |a_{m(0),k}| < 1$. Having chosen the positive integers $m(0), m(1), \ldots, m(j-1)$ and $n(0), n(1), \ldots, n(j-1)$, choose the positive integers $m(j) > m(j-1)$ and $n(j) > n(j-1)$ such that

$$\sum_{k=0}^{\infty} |a_{m(j),k}| > j^2 + 2j + 2; \tag{1.9}$$

$$\sum_{k=0}^{n(j-1)} |a_{m(j),k}| < 1; \tag{1.10}$$

and

$$\sum_{k=n(j)+1}^{\infty} |a_{m(j),k}| < 1. \tag{1.11}$$

Now, define the sequence $x = \{x_k\}$, where

$$x_k = \begin{cases} \frac{|a_{m(j),k}|}{j a_{m(j),k}}, & \text{if } n(j-1) < k \leq n(j),\ a_{m(j),k} \neq 0,\ j = 1, 2, \ldots; \\ 0, & \text{otherwise.} \end{cases}$$

Note that $\lim_{k\to\infty} x_k = 0$. Since A is regular, $\lim_{n\to\infty} (Ax)_n = 0$. However,

$$\begin{aligned}
\left|(Ax)_{m(j)}\right| &= \left|\sum_{k=0}^{\infty} a_{m(j),k} x_k\right| \\
&= \left|\sum_{k=0}^{n(j-1)} a_{m(j),k} x_k + \sum_{k=n(j-1)+1}^{n(j)} a_{m(j),k} x_k + \sum_{k=n(j)+1}^{\infty} a_{m(j),k} x_k\right| \\
&\geq \left|\sum_{k=n(j-1)+1}^{n(j)} a_{m(j),k} x_k\right| - \sum_{k=0}^{n(j-1)} |a_{m(j),k} x_k| - \sum_{k=n(j)+1}^{\infty} |a_{m(j),k} x_k| \\
&> \frac{1}{j} \sum_{k=n(j-1)+1}^{n(j)} |a_{m(j),k}| - 1 - 1, \text{ using (1.10) and (1.11)} \\
&= \frac{1}{j} \left[\sum_{k=0}^{\infty} |a_{m(j),k}| - \sum_{k=0}^{n(j-1)} |a_{m(j),k}| - \sum_{k=n(j)+1}^{\infty} |a_{m(j),k}|\right] - 2 \\
&> \frac{1}{j}[(j^2 + 2j + 2) - 1 - 1] - 2, \text{ using (1.9), (1.10), (1.11)} \\
&= j + 2 - 2
\end{aligned}$$

$$= j, \ j = 1, 2, \ldots.$$

Thus, $\{(Ax)_{m(j)}\}$ diverges, which contradicts the fact that $\{(Ax)_n\}$ converges. Consequently, (1.1) holds. This completes the proof of the theorem. □

Exercise. Prove that $A = (a_{nk})$ is conservative, i.e., $A \in (c, c)$ if and only if (1.1) holds and

$$(i) \lim_{n \to \infty} a_{nk} = \delta_k, \ k = 0, 1, 2, \ldots; \tag{1.12}$$

and

$$(ii) \lim_{n \to \infty} \sum_{k=0}^{\infty} a_{nk} = \delta. \tag{1.13}$$

In such a case, prove that

$$\lim_{n \to \infty} (Ax)_n = s\delta + \sum_{k=0}^{\infty} (x_k - s)\delta_k, \tag{1.14}$$

$\lim_{k \to \infty} x_k = s.$

Definition 1.4 $A = (a_{nk})$ is called a Schur matrix if $A \in (\ell_\infty, c)$, i.e., $\{(Ax)_n\} \in c$, whenever, $x = \{x_k\} \in \ell_\infty$.

The following result gives a characterization of a Schur matrix in terms of the entries of the matrix (see [1–3]).

Theorem 1.2 (Schur) *$A = (a_{nk})$ is a Schur matrix if and only if (1.12) holds and*

$$\sum_{k=0}^{\infty} |a_{nk}| \text{ converges uniformly in } n. \tag{1.15}$$

Proof Sufficiency part. Let (1.2) and (1.15) hold. (1.15) implies that $\sum_{k=0}^{\infty} |a_{nk}|$ converges, $n = 0, 1, 2, \ldots$. Note that (1.12) and (1.15) imply that

$$\sup_{n \geq 0} \sum_{k=0}^{\infty} |a_{nk}| = M < \infty.$$

Thus, for each $r = 0, 1, 2, \ldots$, we have

$$\lim_{n \to \infty} \sum_{k=0}^{r} |a_{nk}| \leq M.$$

Hence,

$$\sum_{k=0}^{r} |\delta_k| \le M, \; r = 0, 1, 2, \dots,$$

and so

$$\sum_{k=0}^{\infty} |\delta_k| < \infty.$$

Thus, if $x = \{x_k\} \in \ell_\infty$, it follows that $\sum_{k=0}^{\infty} a_{nk}x_k$ converges absolutely and uniformly in n. Consequently,

$$\begin{aligned}\lim_{n\to\infty} (Ax)_n &= \lim_{n\to\infty} \sum_{k=0}^{\infty} a_{nk}x_k \\ &= \sum_{k=0}^{\infty} \delta_k x_k,\end{aligned}$$

proving that $\{(Ax)_n\} \in c$, i.e., $A \in (\ell_\infty, c)$, proving the sufficiency part.

Necessity part. Let $A = (a_{nk}) \in (\ell_\infty, c)$. Then, $A \in (c, c)$ and so (1.12) holds. Again, since $A \in (c, c)$, (1.1) holds,

$$\text{i.e., } \sup_{n\ge 0} \sum_{k=0}^{\infty} |a_{nk}| < \infty.$$

As in the sufficiency part of the present theorem, it follows that $\sum_{k=0}^{\infty} |\delta_k| < \infty$. We write

$$b_{nk} = a_{nk} - \delta_k, \; n, k = 0, 1, 2, \dots.$$

Then, $\left\{\sum_{k=0}^{\infty} b_{nk}x_k\right\}_{n=0}^{\infty}$ converges for all $x = \{x_k\} \in \ell_\infty$. We now claim that

$$\sum_{k=0}^{\infty} |b_{nk}| \to 0, \; n \to \infty. \tag{1.16}$$

Suppose not. Then,

$$\overline{\lim_{n\to\infty}} \sum_{k=0}^{\infty} |b_{nk}| = h > 0.$$

So,

$$\sum_{k=0}^{\infty} |b_{mk}| \to h, \ m \to \infty$$

through some subsequence of positive integers. We also note that

$$\lim_{m\to\infty} b_{mk} = 0, \ k = 0, 1, 2, \ldots.$$

We can now find a positive integer $m(1)$ such that

$$\left| \sum_{k=0}^{\infty} |b_{m(1),k}| - h \right| < \frac{h}{10}$$

and

$$|b_{m(1),0}| + |b_{m(1),1}| < \frac{h}{10}.$$

Since $\sum_{k=0}^{\infty} |b_{m(1),k}| < \infty$, we can choose $k(2) > 1$ such that

$$\sum_{k=k(2)+1}^{\infty} |b_{m(1),k}| < \frac{h}{10}.$$

It now follows that

$$\begin{aligned}\left| \sum_{k=2}^{k(2)} |b_{m(1),k}| - h \right| &= \left| \left(\sum_{k=0}^{\infty} |b_{m(1),k}| - h \right) - \left(|b_{m(1),0}| + |b_{m(1),1}| \right) \right. \\ &\quad \left. - \sum_{k=k(2)+1}^{\infty} |b_{m(1),k}| \right| \\ &< \frac{h}{10} + \frac{h}{10} + \frac{h}{10} \\ &= \frac{3h}{10}.\end{aligned}$$

Now choose a positive integer $m(2) > m(1)$ such that

$$\left| \sum_{k=0}^{\infty} |b_{m(2),k}| - h \right| < \frac{h}{10}$$

and

$$\sum_{k=0}^{k(2)} |b_{m(2),k}| < \frac{h}{10}.$$

Then, choose a positive integer $k(3) > k(2)$ such that

$$\sum_{k=k(3)+1}^{\infty} |b_{m(2),k}| < \frac{h}{10}.$$

It now follows that

$$\left| \sum_{k=k(2)+1}^{k(3)} |b_{m(2),k}| - h \right| < \frac{3h}{10}.$$

Continuing this way, we find $m(1) < m(2) < \ldots$ and $1 = k(1) < k(2) < k(3) < \ldots$ such that

$$\sum_{k=0}^{k(r)} |b_{m(r),k}| < \frac{h}{10}; \tag{1.17}$$

$$\sum_{k=k(r+1)+1}^{\infty} |b_{m(r),k}| < \frac{h}{10}; \tag{1.18}$$

and

$$\left| \sum_{k=k(r)+1}^{k(r+1)} |b_{m(r),k}| - h \right| < \frac{3h}{10}. \tag{1.19}$$

We now define a sequence $x = \{x_k\}$ as follows: $x_0 = x_1 = 0$ and

$$x_k = (-1)^r sgn\ b_{m(r),k},$$

if $k(r) < k \leq k(r+1)$, $r = 1, 2, \ldots$. Note that $x = \{x_k\} \in \ell_\infty$ and $\|x\| = 1$. Now,

$$\begin{aligned}
\left| \sum_{k=0}^{\infty} b_{m(r),k} x_k - (-1)^r h \right| &= \left| \sum_{k=0}^{k(r)} b_{m(r),k} x_k + \sum_{k=k(r)+1}^{k(r+1)} b_{m(r),k} x_k \right. \\
&\quad \left. + \sum_{k=k(r+1)+1}^{\infty} b_{m(r),k} x_k - (-1)^r h \right| \\
&= \left| \left\{ \sum_{k=k(r)+1}^{k(r+1)} |b_{m(r),k}| - h \right\} (-1)^r \right.
\end{aligned}$$

$$
\left. + \sum_{k=0}^{k(r)} b_{m(r),k} x_k + \sum_{k=k(r+1)+1}^{\infty} b_{m(r),k} x_k \right|
$$

$$
< \frac{3h}{10} + \frac{h}{10} + \frac{h}{10}, \text{ using (1.17), (1.18) and (1.19)}
$$

$$
= \frac{h}{2}.
$$

Consequently, $\left\{ \sum_{k=0}^{\infty} b_{nk} x_k \right\}_{n=0}^{\infty}$ is not a Cauchy sequence and so it is not convergent, which is a contradiction. Thus, (1.16) holds. So, given $\epsilon > 0$, there exists a positive integer n_0 such that

$$
\sum_{k=0}^{\infty} |b_{nk}| < \epsilon, \; n > n_0. \tag{1.20}
$$

Since $\sum_{k=0}^{\infty} |b_{nk}| < \infty$ for $0 \le n \le n_0$, we can find a positive integer M such that

$$
\sum_{k=M}^{\infty} |b_{nk}| < \epsilon, \; 0 \le n \le n_0. \tag{1.21}
$$

In view of (1.20) and (1.21), we have

$$
\sum_{k=M}^{\infty} |b_{nk}| < \epsilon \text{ for all } n = 0, 1, 2, \ldots,
$$

i.e., $\sum_{k=0}^{\infty} |b_{nk}|$ converges uniformly in n. Since $\sum_{k=0}^{\infty} |\delta_k| < \infty$, it follows that $\sum_{k=0}^{\infty} |a_{nk}|$ converges uniformly in n, proving the necessity part. The proof of the theorem is now complete. □

Using Theorems 1.1 and 1.2, we can deduce the following important result.

Theorem 1.3 (Steinhaus) *An infinite matrix cannot be both a regular and a Schur matrix. In other words, given a regular matrix, there exists a bounded, divergent sequence which is not A-summable.*

Proof Let A be a regular matrix. Then, (1.2) and (1.3) hold. Using (1.15),

$$
\lim_{n \to \infty} \sum_{k=0}^{\infty} a_{nk} = \sum_{k=0}^{\infty} \left(\lim_{n \to \infty} a_{nk} \right)
$$

$$
= 0, \text{ using (1.2)},
$$

which contradicts (1.3). This establishes our claim. □

Exercise. Try to prove Steinhaus theorem without using Schur's theorem, i.e., given a regular matrix, construct a bounded, divergent sequence $x = \{x_k\}$ such that $\{(Ax)_n\}$ diverges.

1.3 A Steinhaus Type Theorem

In the context of Steinhaus theorem, the author introduced the sequence space Λ_r, $r \geq 1$ being a fixed integer as follows (see [4]).

Definition 1.5 Λ_r is defined as the set of all sequences $x = \{x_k\} \in \ell_\infty$ such that

$$|x_{k+r} - x_k| \to 0, k \to \infty,$$

$r \geq 1$ being a fixed integer.

It is easily proved that Λ_r is a closed subspace of ℓ_∞ with respect to the norm defined for elements in ℓ_∞.

The following result, improving Steinhaus theorem, was proved in [4] (it is worth noting that a constructive proof was given).

Theorem 1.4

$$(c, c; P) \cap (\Lambda_r - \bigcup_{i=1}^{r-1} \Lambda_i, c) = \phi.$$

Proof Let $A = (a_{nk})$ be a regular matrix. We can now choose two sequences of positive integers $\{n(m)\}$, $\{k(m)\}$ such that if

$$m = 2p, n(m) > n(m-1), k(m) > k(m-1) + (2m-5)r,$$

then

$$\sum_{k=0}^{k(m-1)+(2m-5)r} |a_{n(m),k}| < \frac{1}{16},$$

$$\sum_{k=k(m-1)}^{\infty} |a_{n(m),k}| < \frac{1}{16};$$

and if

$$m = 2p+1, n(m) > n(m-1), k(m) > k(m-1) + (m-2)r,$$

then,

$$\sum_{k=0}^{k(m-1)+(m-2)r} |a_{n(m),k}| < \frac{1}{16},$$

$$\sum_{k=k(m-1)+(m-2)r+1}^{k(m)} |a_{n(m),k}| > \frac{7}{8},$$

$$\sum_{k=k(m)+1}^{\infty} |a_{n(m),k}| < \frac{1}{16}.$$

Define the sequence $x = \{x_k\}$ as follows:
if $k(2p-1) < k \leq k(2p)$, then

$$x_k = \begin{cases} \frac{2p-2}{2p-1}, & k = k(2p-1)+1, \\ 1, & k(2p-1)+1 < k \leq k(2p-1)+r, \\ \frac{2p-3}{2p-1}, & k = k(2p-1)+r+1, \\ 1, & k(2p-1)+r+1 < k \leq k(2p-1)+2r, \\ \vdots \\ 1, & k(2p-1)+(2p-4)r+1 < k \leq k(2p-1)+(2p-3)r, \\ \frac{1}{2p-1}, & k = k(2p-1)+(2p-3)r+1, \\ \frac{2p-2}{2p-1}, & k(2p-1)+(2p-3)r+1 < k \leq k(2p-1)+(2p-2)r, \\ 0, & k = k(2p-1)+(2p-2)r+1, \\ \vdots \\ \frac{1}{2p-1}, & k(2p-1)+(4p-6)r+1 < k \leq k(2p-1)+(4p-5)r, \\ 0, & k(2p-1)+(4p-5)r < k \leq k(2p), \end{cases}$$

and if $k(2p) < k \leq k(2p+1)$, then

$$x_k = \begin{cases} \frac{1}{2p}, & k(2p) < k \leq k(2p)+r, \\ \frac{2}{2p}, & k(2p)+r < k \leq k(2p)+2r, \\ \vdots \\ \frac{2p-1}{2p}, & k(2p)+(2p-2)r < k \leq k(2p)+(2p-1)r, \\ 1, & k(2p)+(2p-1)r < k \leq k(2p+1). \end{cases}$$

Note that,

$$\text{if } k(2p-1) < k \le k(2p),$$
$$|x_{k+r} - x_k| < \frac{1}{2p-1},$$

while

$$\text{if } k(2p) < k \le k(2p+1),$$
$$|x_{k+r} - x_k| < \frac{1}{2p}.$$

Thus,

$$|x_{k+r} - x_k| \to 0, \quad k \to \infty$$

and so $x = \{x_k\} \in \Lambda_r$.
However,

$$|x_{k+1} - x_k| = \frac{2p-2}{2p-1}, \quad \text{if } k = k(2p-1) + (2p-3)r, \quad p = 1, 2, \ldots.$$

Hence,

$$|x_{k+1} - x_k| \not\to 0, \quad k \to \infty$$

and consequently $x = \{x_k\} \notin \Lambda_1$. In a similar fashion, we can prove that

$$x \notin \Lambda_i, \quad i = 2, 3, \ldots, (r-1).$$

Thus,

$$x \notin \bigcup_{i=1}^{r-1} \Lambda_i$$

and so

$$x \in \Lambda_r - \bigcup_{i=1}^{r-1} \Lambda_i.$$

Further,

$$\left.\begin{array}{l} |(Ax)_{n(2p)}| < \frac{1}{16} + \frac{1}{16} = \frac{1}{8}, \\ |(Ax)_{n(2p+1)}| > \frac{7}{8} - \frac{1}{16} - \frac{1}{16} = \frac{3}{4} \end{array}\right\}, \quad p = 1, 2, \ldots,$$

which shows that $\{(Ax)_n\} \notin c$, completing the proof of the theorem. □

Remark 1.1 We note that $(c, c; P) \cap (\Lambda_r, c) = \phi$. Since $(\ell_\infty, c) \subseteq (\Lambda_r, c)$, it follows that $(c, c; P) \cap (\ell_\infty, c) = \phi$, which is Steinhaus theorem.

We call such results Steinhaus type theorems. More explicitly, whenever there is some notion of limit or sum in the sequence spaces X, Y, we denote by $(X, Y; P)$ that subclass of (X, Y) consisting of all infinite matrices which preserve this limit or sum. Then, results of the form,

$$(X, Y; P) \cap (Z, Y) = \phi,$$

$X \subsetneqq Z$, are called Steinhaus type theorems.

1.4 The Role Played by the Sequence Spaces Λ_r

Let us now study in detail the role played by the sequence spaces Λ_r. It is well known that an infinite matrix which sums all sequences of 0's and 1's sums all bounded sequences (see [5]). It is clear that any Cauchy sequence is in $\bigcap_{r=1}^{\infty} \Lambda_r$ so that each Λ_r is a sequence space containing the space $\mathscr{C}$ of Cauchy sequences. It is to be noted that

$$\mathscr{C} \subsetneqq \bigcap_{r=1}^{\infty} \Lambda_r.$$

Though Λ_r do not form a tower between $\mathscr{C}$ and ℓ_∞, they can be deemed to reflect the measure of non-Cauchy nature of sequences contained in them. It is also easy to prove that $\Lambda_r \subseteq \Lambda_s$ if and only if s is a multiple of r and that $\Lambda_r \cap \Lambda_{r+1} = \Lambda_1$. It is worthwhile to observe the nature of location of sequences of 0's and 1's in these spaces Λ_r. In the first instance, we note that a sequence of 0's and 1's is in Λ_r if and only if it is periodic with period r eventually. Consequently, any sequence of 0's and 1's is in $\ell_\infty - \bigcup_{r=1}^{\infty} \Lambda_r$ if and only if it is non-periodic. Let $\mathcal{NP}$ denote the set of all sequences of 0's and 1's in $\ell_\infty - \bigcup_{r=1}^{\infty} \Lambda_r$, i.e., the set of all sequences of 0's and 1's which are non-periodic. The following sequences

$$\begin{aligned} e_i^{(r)} &= \{e_{ik}^{(r)}\}_{k=0}^{\infty} \\ &= \left\{\underbrace{1, 1, \ldots, 1}_{i}, \underbrace{0, 0, \ldots, 0}_{r-i}, \underbrace{1, 1, \ldots, 1}_{i}, \underbrace{0, 0, \ldots, 0}_{r-i}, \ldots\right\}, \quad i = 1, 2, \ldots, r, \end{aligned} \tag{1.22}$$

have a role to play in the structure of Λ_r. Note that there are sequences of 0's and 1's in Λ_r which are not necessarily of the form (1.22),

$$\text{e.g.,} \quad \left\{1, 0, 1, \underbrace{0, 0, \ldots, 0}_{r-3}, 1, 0, 1, \underbrace{0, 0, \ldots, 0}_{r-3}, \ldots\right\}.$$

We also note that $A = (a_{nk})$ sums the sequences in (1.22) if and only if

$$\lim_{n\to\infty}\left(\sum_{k=0}^{\infty} a_{n,j+kr}\right) \text{ exists, } j = 0, 1, 2, \ldots, r-1. \tag{1.23}$$

The role of the sequences in (1.22) is illustrated by the following theorem (see [6], Theorem 1.1).

Theorem 1.5 *$A = (a_{nk})$ sums every sequence of 0's and 1's in Λ_r if and only if (1.1) and (1.23) hold.*

In view of Schur's version of Steinhaus theorem [5], viz. given a regular matrix A, there exists a sequence of 0's and 1's not summable A, the following questions arise:

(1) Given a regular matrix A, does there exist a sequence of 0's and 1's in $\ell_\infty - \bigcup_{r=1}^{\infty}\Lambda_r$, i.e., non-periodic which is not summable A?

(2) Given a regular matrix A, does there exist a sequence of 0's and 1's in $\bigcup_{r=1}^{\infty}\Lambda_r$, i.e., eventually periodic which is not summable A?

The following example provides the negative answer to question (2). Consider the infinite matrix

$$A = \begin{pmatrix} 1 & 0 & 0 & 0 & 0 & 0 & \ldots \\ 0 & \frac{1}{2} & \frac{1}{2} & 0 & 0 & 0 & \ldots \\ 0 & 0 & \frac{1}{3} & \frac{1}{3} & \frac{1}{3} & 0 & \ldots \\ \ldots & \ldots & \ldots & \ldots & \ldots & \ldots & \ldots \end{pmatrix}.$$

We can prove that A sums all sequences of 0's and 1's in $\bigcup_{r=1}^{\infty}\Lambda_r$, using Theorem 7 of [7], noting that an eventually periodic sequence of 0's and 1's is almost convergent.

We now record some of the structural properties of ℓ_∞, vis-a-vis, the set of all sequences of 0's and 1's. It can be proved that the closed linear span of the set of all sequences of 0's and 1's in the supremum norm, i.e., $\|x\| = \sup_{k\geq 0}|x_k|$, $x = \{x_k\} \in \ell_\infty$ is ℓ_∞ (see [8] for details). The following result is an improvement of this assertion.

Theorem 1.6 *The closed linear span of $\mathcal{NP}$ is ℓ_∞.*

Proof It suffices to show that periodic sequences of 0's and 1's, which are periodic from the beginning, are in the closed span of $\mathcal{NP}$. In fact, they are in the linear span of $\mathcal{NP}$. To this end, we show that any such sequence is the difference of two sequences of $\mathcal{NP}$. It is clear that any sequence of 0's and 1's which converges to 0 or 1 can be expressed as the difference of two sequences of $\mathcal{NP}$. Hence, we shall take a divergent sequence $x = \{x_k\}$, $x_{k+r} = x_k$, $k = 0, 1, 2, \ldots$. Let $\{k(i)\}$ be a strictly increasing sequence of positive integers such that $k(i+1) - k(i) \to \infty$, $i \to \infty$, $x_{k(i)} = 1$, $x_{k(i)+1} = 0$, $i = 1, 2, \ldots$. We now construct two sequences $x^{(1)} = \{x_k^{(1)}\}$, $x^{(2)} = \{x_k^{(2)}\}$, using x as follows:

$$x_k^{(1)} \begin{array}{l} = 1, \ \ k = k(i) + 1; \\ = x_k, \ k \neq k(i) + 1 \end{array} \Bigg\}, \quad i = 1, 2, \ldots;$$

$$x_k^{(2)} \begin{array}{l} = 1, \ k = k(i) + 1; \\ = 0, \ k \neq k(i) + 1 \end{array} \Bigg\}, \quad i = 1, 2, \ldots.$$

By construction, $x = x^{(1)} - x^{(2)}$. It is easy to observe that $x^{(2)}$ is non-periodic. To prove that $x^{(1)}$ is non-periodic, suppose $x^{(1)}$ has period p. Since both $x^{(1)}$ and x have period pr, $x^{(2)} = x^{(1)} - x$ has period pr, a contradiction. This completes the proof of the theorem. □

The following is an important result due to Schur [5].

Theorem 1.7 *Any matrix A which sums all sequences of 0's and 1's is necessarily a Schur matrix, i.e., A sums all bounded sequences.*

In view of Theorem 1.6, we have an improvement of Theorem 1.7.

Theorem 1.8 *Any matrix A which sums all sequences in $\mathcal{NP}$ is necessarily a Schur matrix.*

Theorem 1.8 provides an affirmative answer to question (1) already recorded and is an improvement of Steinhaus theorem.

Thus, it turns out that the probability of success in our search for a sequence of 0's and 1's not summable by a given regular matrix A is more when we concentrate on non-periodic than on eventually periodic sequences.

Remark 1.2 In the context of Theorem 1.4, one may enquire whether a matrix which sums all sequences of 0's and 1's in $\bigcup_{r=1}^{\infty} \Lambda_r$ sums all sequences in $\bigcup_{r=1}^{\infty} \Lambda_r$. The answer to this query is, however, in the negative. For, if $x = \{x_k\}$ is a sequence of 0's and 1's in $\bigcup_{r=1}^{\infty} \Lambda_r$, then $x_{k+r} = x_k$, $k \geq k_0$ and for some integer $r \geq 1$, k_0 being a positive integer, i.e., x is eventually periodic and hence $(C, 1)$ summable as is seen directly or using the idea of almost convergence (see [9, p. 12]). The sequence

$$\left\{0, 1, 0, \frac{1}{2}, \frac{2}{2}, \frac{1}{2}, 0, \frac{1}{4}, \frac{2}{4}, \frac{3}{4}, \frac{4}{4}, \frac{3}{4}, \frac{2}{4}, \frac{1}{4}, 0, \ldots\right\}$$

is in Λ_1 but it is not $(C, 1)$ summable.

In the context of $\left\{\overline{\bigcup_{r=1}^{\infty} \Lambda_r}\right\}$, i.e., the closure of $\bigcup_{r=1}^{\infty} \Lambda_r$ in ℓ_∞, we introduce the notion of "generalized semiperiodic sequences" (for the definition of "semiperiodic sequences," one can refer to [10]).

Definition 1.6 $x = \{x_k\}$ is called a "generalized semiperiodic sequence," if for any $\epsilon > 0$, there exist positive integers n, k_0 such that

$$|x_k - x_{k+sn}| < \epsilon, \quad k \geq k_0, s = 0, 1, 2, \ldots.$$

Let Q denote the set of all generalized semiperiodic sequences. One can prove that Q is a closed linear subspace of ℓ_∞. Further,

$$Q \subseteq \left\{\overline{\bigcup_{r=1}^{\infty} \Lambda_r}\right\}.$$

1.5 More Steinhaus Type Theorems

In this section, we prove some more Steinhaus type theorems.

The sequence spaces ℓ_p, $p \geq 1$, c_0 are defined as usual:

$$\ell_p = \{x = \{x_k\} : \sum_{k=0}^{\infty} |x_k|^p < \infty\}, p \geq 1;$$

$$c_0 = \{x = \{x_k\} : \lim_{k\to\infty} x_k = 0\}.$$

Note that $\ell_p \subset c_0 \subset c \subset \ell_\infty$, where $p \geq 1$. Just for convenience, we write $\ell_1 = \ell$. Let $(\ell, c; P')$ denote the class of all infinite matrices $A \in (\ell, c)$ such that $\lim_{n\to\infty} (Ax)_n = \sum_{k=0}^{\infty} x_k$, $x = \{x_k\} \in \ell$. The following result is known ([11], p. 4, 17).

Theorem 1.9 *$A = (a_{nk}) \in (\ell, c)$ if and only if*

$$\sup_{n,k} |a_{nk}| < \infty, \tag{1.24}$$

and (1.1) holds.

We now prove the following (see [12, Theorem 2.1]).

Theorem 1.10 *$A = (a_{nk}) \in (\ell, c; P')$ if and only if (1.24), (1.1) hold with $\delta_k = 1$, $k = 0, 1, 2, \ldots$.*

Proof Let $A \in (\ell, c; P')$. Then, (1.24) holds. Let e_k be the sequence in which 1 occurs in the kth place and 0 elsewhere,
i.e.,

$$e_k = \{x_i^{(k)}\}_{i=0}^{\infty},$$

where

$$x_i^{(k)} = \begin{cases} 1, & i = k; \\ 0, & \text{otherwise}, \end{cases}$$

$k = 0, 1, 2, \ldots$. Then $e_k \in \ell$, $k = 0, 1, 2, \ldots$ and $\sum_{i=0}^{\infty} x_i^{(k)} = 1$, $k = 0, 1, 2, \ldots$. Now, $(Ae_k)_n = a_{nk}$ so that $\lim_{n\to\infty} a_{nk} = 1, k = 0, 1, 2, \ldots$, i.e., $\delta_k = 1, k = 0, 1, 2, \ldots$. Conversely, let (1.24), (1.1) hold and $\delta_k = 1$, $k = 0, 1, 2, \ldots$. Let $x = \{x_k\} \in \ell$. In view of (1.24), $(Ax)_n$ is defined, $n = 0, 1, 2, \ldots$. Now,

$$\begin{aligned}(Ax)_n &= \sum_{k=0}^{\infty} a_{nk}x_k \\ &= \sum_{k=0}^{\infty}(a_{nk} - 1)x_k + \sum_{k=0}^{\infty} x_k,\end{aligned}$$

this being true since $\sum_{k=0}^{\infty} a_{nk}x_k$ and $\sum_{k=0}^{\infty} x_k$ both converge. Since $\sum_{k=0}^{\infty} |x_k| < \infty$, given $\epsilon > 0$, there exists a positive integer N such that

$$\sum_{k=N+1}^{\infty} |x_k| < \frac{\epsilon}{2A}, \tag{1.25}$$

where $A = \sup_{n,k} |a_{nk} - 1|$. Since $\lim_{n\to\infty} a_{nk} = 1$, $k = 0, 1, 2, \ldots, N$, we can choose a positive integer $N' > N$ such that

$$|a_{nk} - 1| < \frac{\epsilon}{2(N+1)M}, n \geq N', k = 0, 1, 2, \ldots, N, \tag{1.26}$$

where $M > 0$ is such that $|x_k| \leq M$, $k = 0, 1, 2, \ldots$. Now, for $n \geq N'$,

$$\left|\sum_{k=0}^{\infty}(a_{nk}-1)x_k\right| \leq \sum_{k=0}^{N}|a_{nk}-1||x_k| + \sum_{k=N+1}^{\infty}|a_{nk}-1||x_k|$$
$$< (N+1)\frac{\epsilon}{2(N+1)M}M + A\frac{\epsilon}{2A},$$
$$\text{in view of (1.25) and (1.26)}$$
$$= \epsilon,$$

so that

$$\lim_{n\to\infty}\sum_{k=0}^{\infty}(a_{nk}-1)x_k = 0.$$

Consequently,

$$\lim_{n\to\infty}(Ax)_n = \sum_{k=0}^{\infty}x_k$$

and so $A \in (\ell, c; P')$, completing the proof of the theorem. □

Using Theorem 1.10, we have the following Steinhaus type theorem (see [12, Theorem 2.2]).

Theorem 1.11

$$(\ell, c; P') \cap (\ell_p, c) = \phi,\ p > 1.$$

Proof Let $A = (a_{nk}) \in (\ell, c; P') \cap (\ell_p, c)$, $p > 1$. It is known ([11], p. 4, 16) that $A \in (\ell_p, c)$, $p > 1$, if and only if (1.1) holds and

$$\sup_{n\geq 0}\sum_{k=0}^{\infty}|a_{nk}|^q < \infty, \tag{1.27}$$

where $\frac{1}{p}+\frac{1}{q}=1$. It now follows that $\sum_{k=0}^{\infty}|\delta_k|^q < \infty$, which contradicts the fact that $\delta_k = 1$, $k = 0, 1, 2, \ldots$, since $A \in (\ell, c; P')$ and consequently $\sum_{k=0}^{\infty}|\delta_k|^q$ diverges. This proves our claim. □

Remark 1.3 Since $\ell_p \subset c_0 \subset c \subset \ell_\infty$,

$$(\ell_\infty, c) \subset (c, c) \subset (c_0, c) \subset (\ell_p, c),\ p > 1.$$

So, we have

$$(\ell, c; P') \cap (X, c) = \phi,$$

when

$$X = \ell_\infty, c, c_0, \ell_p, p > 1.$$

Remark 1.4 ℓ_∞ is a Banach space with respect to the norm $\|x\| = \sup_{k \geq 0} |x_k|$, $x = \{x_k\} \in \ell_\infty$. c_0 and c are closed subspaces of ℓ_∞. ℓ_p, $p > 1$, is a Banach space with respect to the norm

$$\|x\| = \left(\sum_{k=0}^{\infty} |x_k|^p \right)^{\frac{1}{p}},$$

$x = \{x_k\} \in \ell_p$. Let $BL(\ell, c)$ denote the space of all bounded linear mappings of ℓ into c. We note that if $A \in (\ell, c; P')$, then A is bounded and $\|A\| = \sup_{n,k} |a_{nk}|$. However, $(\ell, c; P')$ is not a subspace of $BL(\ell, c)$, since $\lim_{n \to \infty} 2a_{nk} = 2$, $k = 0, 1, 2, \ldots$ and consequently $2A \notin (\ell, c; P')$ when $A \in (\ell, c; P')$.

Let

$$cs = \left\{ x = \{x_k\} : \sum_{k=0}^{\infty} x_k \text{ converges} \right\}.$$

Note that $\ell \subset cs \subset c_0 \subset c \subset \ell_\infty$. We write $A = (a_{nk}) \in (\ell, \ell; P)$ if $A \in (\ell, \ell)$ and $\sum_{n=0}^{\infty} (Ax)_n = \sum_{k=0}^{\infty} x_k, x = \{x_k\} \in \ell$. We also write $A \in (\ell, \ell; P)'$ if $A \in (\ell, \ell; P)$ with

$$\lim_{k \to \infty} a_{nk} = 0, \quad n = 0, 1, 2, \ldots. \tag{1.28}$$

It is well known (see, for instance, [2, p. 189]) that $A \in (\ell, \ell; P)$ if and only if $A \in (\ell, \ell)$ and

$$\sum_{n=0}^{\infty} a_{nk} = 1, \quad k = 0, 1, 2, \ldots. \tag{1.29}$$

Maddox [13] noted that

$$(\ell, \ell; P) \cap (cs, \ell) \neq \phi.$$

We now prove that given a $(\ell, \ell; P)'$ matrix A, there exists a sequence $x = \{x_k\} \in cs$ whose A-transform is not in ℓ. In particular, a lower triangular $(\ell, \ell; P)$ matrix cannot belong to the class (cs, ℓ). We need the following lemma, the proof of which is modeled on that of Fridy's (see [14]).

Lemma 1.1 (see [15, pp. 140–142, Lemma]) *If* $A = (a_{nk}) \in (\ell, \ell)$ *and satisfies (1.28) and if*

$$\overline{\lim_{k\to\infty}}\left|\sum_{n=0}^{\infty} a_{nk}\right| > 0, \tag{1.30}$$

then there exists a sequence $x = \{x_k\} \in cs$, $Ax = \{(Ax)_n\} \notin \ell$.

Proof By hypothesis, for some $\epsilon > 0$, there exists a strictly increasing sequence $\{k(i)\}$ of positive integers such that

$$\left|\sum_{n=0}^{\infty} a_{n,k(i)}\right| \geq 2\epsilon, \quad i = 1, 2, \ldots.$$

In particular,

$$\left|\sum_{n=0}^{\infty} a_{n,k(1)}\right| \geq 2\epsilon.$$

We then choose a positive integer $n(1)$ such that

$$\sum_{n>n(1)} |a_{n,k(1)}| < \min\left(\frac{1}{2}, \frac{\epsilon}{2}\right),$$

this being possible since $\sum_{n=0}^{\infty} |a_{nk}| < \infty$, $k = 0, 1, 2, \ldots$, in view of the fact that $A \in (\ell, \ell)$. Now, it follows that

$$\left|\sum_{n=0}^{n(1)} a_{n,k(1)}\right| > \epsilon.$$

In general, having chosen $k(j), n(j)$, $j \leq m - 1$, choose a positive integer $k(m) > k(m-1)$ such that

$$\left|\sum_{n=0}^{\infty} a_{n,k(m)}\right| \geq 2\epsilon,$$

$$\sum_{n=0}^{n(m-1)} |a_{n,k(m)}| < \min\left(\frac{1}{2}, \frac{\epsilon}{2}\right),$$

and then choose a positive integer $n(m) > n(m-1)$ such that

$$\sum_{n>n(m)} |a_{n,k(m)}| < \min\left(\frac{1}{2^m}, \frac{\epsilon}{2}\right),$$

so that

$$\left|\sum_{n=n(m-1)+1}^{n(m)} a_{n,k(m)}\right| > 2\epsilon - \frac{\epsilon}{2} - \frac{\epsilon}{2}$$
$$= \epsilon.$$

Let the sequence $x = \{x_k\}$ be defined by

$$x_k = \begin{cases} \frac{(-1)^{i+1}}{i}, & k = k(i); \\ 0, & k \neq k(i), i = 1, 2, \ldots. \end{cases}$$

It is clear that $x = \{x_k\} \in cs$. Defining $n(0) = 0$, we have

$$\begin{aligned}
\sum_{n=0}^{n(N)} |(Ax)_n| &\geq \sum_{m=1}^{N} \sum_{n=n(m-1)+1}^{n(m)} |(Ax)_n| \\
&= \sum_{m=1}^{N} \sum_{n=n(m-1)+1}^{n(m)} \left|\sum_{i=1}^{\infty} a_{n,k(i)} x_{k(i)}\right| \\
&= \sum_{m=1}^{N} \sum_{n=n(m-1)+1}^{n(m)} \left|\sum_{i=1}^{\infty} a_{n,k(i)} \frac{(-1)^{i+1}}{i}\right| \\
&\geq \sum_{m=1}^{N} \sum_{n=n(m-1)+1}^{n(m)} \left\{\left|\frac{(-1)^{m+1}}{m} a_{n,k(m)}\right| - \sum_{\substack{i=1 \\ i\neq m}}^{\infty} \left|\frac{(-1)^{i+1}}{i} a_{n,k(i)}\right|\right\} \\
&= \sum_{m=1}^{N} \sum_{n=n(m-1)+1}^{n(m)} \left\{\frac{1}{m} |a_{n,k(m)}| - \sum_{\substack{i=1 \\ i\neq m}}^{\infty} \frac{1}{i} |a_{n,k(i)}|\right\} \\
&> \epsilon \sum_{m=1}^{N} \frac{1}{m} - \sum_{m=1}^{N} \sum_{n=n(m-1)+1}^{n(m)} \sum_{\substack{i=1 \\ i\neq m}}^{\infty} |a_{n,k(i)}|, \text{ since } \frac{1}{i} \leq 1.
\end{aligned}$$

Now,

$$\sum_{m=1}^{\infty} \sum_{n=n(m-1)+1}^{n(m)} \sum_{i<m} |a_{n,k(i)}| \leq \sum_{m=1}^{\infty} \frac{1}{2^m}$$
$$= 1.$$

Similarly,

$$\sum_{m=1}^{\infty} \sum_{n=n(m-1)+1}^{n(m)} \sum_{i>m} |a_{n,k(i)}| \leq \frac{1}{2}.$$

So,

$$\sum_{n=0}^{n(N)} |(Ax)_n| > \epsilon \sum_{m=1}^{N} \frac{1}{m} - \frac{3}{2}.$$

Since $\sum_{m=1}^{\infty} \frac{1}{m}$ diverges, $Ax \notin \ell$, completing the proof of the lemma. □

We now prove a Steinhaus type result.

Theorem 1.12 ([15, p. 143, Theorem])

$$(\ell, \ell; P)' \cap (cs, \ell) = \phi.$$

Proof Let $A = (a_{nk}) \in (\ell, \ell; P)' \cap (cs, \ell)$. Since $\sum_{n=0}^{\infty} a_{nk} = 1, k = 0, 1, 2, \ldots,$

$$\overline{\lim_{k\to\infty}} \left| \sum_{n=0}^{\infty} a_{nk} \right| = 1 > 0.$$

In view of the preceding Lemma 1.1, there exists a sequence $x = \{x_k\} \in cs$ such that $Ax \notin \ell$, which is a contradiction. This completes the proof. □

Remark 1.5 If the condition (1.28) is dropped, the above theorem fails to hold as the following example illustrates. The infinite matrix

$$A = (a_{nk}) = \begin{pmatrix} 1 & 1 & 1 & \cdots \\ 0 & 0 & 0 & \cdots \\ 0 & 0 & 0 & \cdots \\ \cdots & \cdots & \cdots & \cdots \end{pmatrix}$$

is in $(\ell, \ell; P) \cap (cs, \ell)$ but $a_{0k} \not\to 0, k \to \infty$.

For more Steinhaus type theorems, the reader can refer to [16–19].

References

1. Hardy, G.H.: Divergent Series. Oxford University Press, Oxford (1949)
2. Maddox, I.J.: Elements of Functional Analysis. Cambridge University Press, Cambridge (1970)

3. Peyerimhoff, A.: Lectures on Summability. Lecture Notes in Mathematics, vol. 107. Springer, Berlin (1969)
4. Natarajan, P.N.: A Steinhaus type theorem. Proc. Amer. Math. Soc. **99**, 559–562 (1987)
5. Schur, I.: Uber lineare Transformationen in der Theorie der unendlichen Reihen. J. Reine Angew. Math. **151**, 79–111 (1921)
6. Natarajan, P.N.: On certain spaces containing the space of Cauchy sequences. J. Orissa Math. Soc. **9**, 1–9 (1990)
7. Lorentz, G.G.: A contribution to the theory of divergent sequences. Acta Math. **80**, 167–190 (1948)
8. Hill, J.D., Hamilton, H.J.: Operation theory and multiple sequence transformations. Duke Math. J. **8**, 154–162 (1941)
9. Zeller, K., Beekmann, W.: Theorie der Limitierungverfahren. Springer, Berlin (1970)
10. Berg, I.D., Wilansky, A.: Periodic, almost periodic and semiperiodic sequences. Michigan Math. J. **9**, 363–368 (1962)
11. Stieglitz, M., Tietz, H.: Matrixtransformationen von Folgenraümen Eine Ergebnisübersicht. Math. Z. **154**, 1–16 (1977)
12. Natarajan, P.N.: Some Steinhaus type theorems over valued fields. Ann. Math. Blaise Pascal **3**, 183–188 (1996)
13. Maddox, I.J.: On theorems of Steinhaus type. J. London Math. Soc. **42**, 239–244 (1967)
14. Fridy, J.A.: Properties of absolute summability matrices. Proc. Amer. Math. Soc. **24**, 583–585 (1970)
15. Natarajan, P.N.: A theorem of Steinhaus type. J. Anal. **5**, 139–143 (1997)
16. Natarajan, P.N.: Some more Steinhaus type theorems over valued fields. Ann. Math. Blaise Pascal **6**, 47–54 (1999)
17. Natarajan, P.N.: Some more Steinhaus type theorems over valued fields II. Commun. Math. Anal. **5**, 1–7 (2008)
18. Natarajan, P.N.: Steinhaus type theorems for $(C, 1)$ summable sequences. Comment Math. Prace Mat. **54**, 21–27 (2014)
19. Natarajan, P.N.: Steinhaus type theorems for summability matrices. Adv. Dev. Math. Sci. **6**, 1–8 (2014)

Chapter 2
Core of a Sequence and the Matrix Class (ℓ, ℓ)

In this chapter, we present a study of the core of a sequence and properties of the matrix class (ℓ, ℓ). This chapter is divided into 4 sections. In the first section, we introduce the core of a sequence and study some of its properties. In the second section, we prove an improvement of Sherbakoff's result. This result leads to a short and elegant proof of Knopp's core theorem. The third section is devoted to a study of the matrix class (ℓ, ℓ) in the context of a convolution product $*$. In the final section, we prove a Mercerian theorem for the Banach algebra (ℓ, ℓ) under the convolution product $*$.

2.1 Core of a Sequence

The core of a complex sequence is defined as follows.

Definition 2.1 If $x = \{x_k\}$ is a complex sequence, we denote by $K_n(x)$, $n = 0, 1, 2, \ldots$, the smallest closed convex set containing $x_n, x_{n+1}, \ldots$.

$$\mathscr{K}(x) = \bigcap_{n=0}^{\infty} K_n(x)$$

is defined as the core of x.

It is known [1] that if $x = \{x_k\}$ is bounded,

$$\mathscr{K}(x) = \bigcap_{z \in \mathbb{C}} C_{\overline{\lim\limits_{n\to\infty}} |z - x_n|}(z),$$

where $C_r(z)$ is the closed ball centered at z and radius r. Sherbakhoff [1] generalized the notion of the core of a bounded complex sequence by introducing the idea of the generalized α-core $\mathscr{K}^{(\alpha)}(x)$ of a bounded complex sequence as

P.N. Natarajan, *Classical Summability Theory*, DOI 10.1007/978-981-10-4205-8_2

$$\mathscr{K}^{(\alpha)}(x) = \bigcap_{z \in \mathbb{C}} C_{\alpha \overline{\lim_{n\to\infty}} |z - x_n|}(z), \quad \alpha \geq 1. \tag{2.1}$$

When $\alpha = 1$, $\mathscr{K}^{(\alpha)}(x)$ reduces to the usual core $\mathscr{K}(x)$. Sherbakhoff [1] showed that under the condition

$$\overline{\lim_{n\to\infty}} \left(\sum_{k=0}^{\infty} |a_{nk}| \right) = \alpha, \quad \alpha \geq 1, \tag{2.2}$$

$$\mathscr{K}(A(x)) \subseteq \mathscr{K}^{(\alpha)}(x).$$

Natarajan [2] improved Sherbakhoff's result by showing that his result works with the less stringent precise condition

$$\overline{\lim_{n\to\infty}} \left(\sum_{k=0}^{\infty} |a_{nk}| \right) \leq \alpha, \quad \alpha \geq 1, \tag{2.3}$$

(2.3) being also necessary besides the regularity of A for

$$\mathscr{K}(A(x)) \subseteq \mathscr{K}^{(\alpha)}(x),$$

for any bounded complex sequence x. This result for the case $\alpha = 1$ yields a simple and very elegant proof of Knopp's core theorem (see, for instance, [3]).

2.2 Natarajan's Theorem and Knopp's Core Theorem

Natarajan's theorem is

Theorem 2.1 ([2, Theorem 2.1]) $A = (a_{nk})$ *is such that*

$$\mathscr{K}(A(x)) \subseteq \mathscr{K}^{(\alpha)}(x), \quad \alpha \geq 1,$$

for any bounded sequence x if and only if A is regular and satisfies (2.3),

$$i.e., \overline{\lim_{n\to\infty}} \left(\sum_{k=0}^{\infty} |a_{nk}| \right) \leq \alpha, \quad \alpha \geq 1.$$

Proof Let $x = \{x_n\}$ be a bounded sequence. If $y \in \mathscr{K}(A(x))$, for any z,

$$|y - z| \leq \overline{\lim_{n\to\infty}} |z - (Ax)_n|.$$

If A is a regular matrix satisfying (2.3), then

$$\begin{aligned}|y - z| &\leq \overline{\lim_{n\to\infty}} |z - (Ax)_n| \\ &= \overline{\lim_{n\to\infty}} \left| \sum_{k=0}^{\infty} a_{nk}(z - x_k) \right| \\ &\leq \alpha \overline{\lim_{k\to\infty}} |z - x_k|,\end{aligned}$$

$$i.e., \ y \in C_{\alpha \overline{\lim}_{k\to\infty} |z - x_k|}(z) \ \text{ for any } z,$$

which implies that

$$\mathscr{K}(A(x)) \subseteq \mathscr{K}^{(\alpha)}(x).$$

Conversely, let

$$\mathscr{K}(A(x)) \subseteq \mathscr{K}^{(\alpha)}(x).$$

Then, it is clear that A is regular by considering convergent sequences for which

$$\mathscr{K}^{(\alpha)}(x) = \left\{ \lim_{n\to\infty} x_n \right\}.$$

It remains to prove (2.3). Let, if possible,

$$\overline{\lim_{n\to\infty}} \left(\sum_{k=0}^{\infty} |a_{nk}| \right) > \alpha.$$

Then,

$$\overline{\lim_{n\to\infty}} \left(\sum_{k=0}^{\infty} |a_{nk}| \right) = \alpha + h, \text{ for some } h > 0.$$

Using the hypothesis and the fact that A is regular, we can choose two strictly increasing sequences $\{n(i)\}$ and $\{k(n(i))\}$ of positive integers such that

$$\sum_{k=0}^{k(n(i-1))} |a_{n(i),k}| < \frac{h}{8},$$

$$\sum_{k=k(n(i-1))+1}^{k(n(i))} |a_{n(i),k}| > \alpha + \frac{h}{4}$$

and

$$\sum_{k=k(n(i))+1}^{\infty} |a_{n(i),k}| < \frac{h}{8}.$$

Define the sequence $x = \{x_k\}$ by

$$x_k = sgn(a_{n(i),k}),\ k(n(i-1)) \leq k < k(n(i)), i = 1, 2, \ldots.$$

Now,

$$\begin{aligned}
|(Ax)_{n(i)}| &\geq \sum_{k=k(n(i-1))+1}^{k(n(i))} |a_{n(i),k}| - \sum_{k=0}^{k(n(i-1))} |a_{n(i),k}| \\
&- \sum_{k=k(n(i))+1}^{\infty} |a_{n(i),k}| \\
&> \alpha + \frac{h}{4} - \frac{h}{8} - \frac{h}{8} \\
&= \alpha,\ \ i = 1, 2, \ldots.
\end{aligned} \tag{2.4}$$

By the regularity of A, $\{(Ax)_{n(i)}\}_{i=1}^{\infty}$ is a bounded sequence. It has a convergent subsequence whose limit cannot be in $C_\alpha(0)$, in view of (2.4). Using (2.1), we have, $\mathscr{K}^{(\alpha)}(x) \subseteq C_\alpha(0)$ for the sequence x chosen above. This leads to a contradiction of the fact that $\mathscr{K}(A(x)) \subseteq \mathcal{K}^{(\alpha)}(x)$, completing the proof of the theorem. □

Remark 2.1 Condition (2.3) cannot be relaxed if $\mathcal{K}(A(x))$ were to be contained in $\mathcal{K}^{(\alpha)}(x)$. The infinite matrix

$$\begin{pmatrix} 1 & \lambda & -\lambda & 0 & 0 & 0 & \cdots \\ 0 & 1 & \lambda & -\lambda & 0 & 0 & \cdots \\ 0 & 0 & 1 & \lambda & -\lambda & 0 & \cdots \\ \cdots & \cdots & \cdots & \cdots & \cdots & \cdots & \cdots \end{pmatrix},$$

where $|\lambda| > \alpha$, transforms the sequence $\{1, 0, 1, 0, \ldots\}$ into the sequence $\{\lambda, 1 - \lambda, \lambda, 1 - \lambda, \ldots\}$. $\mathcal{K}^{(\alpha)}(x) \subset C_\alpha(0)$ while $\lambda \in \mathscr{K}(A(x))$ and $\lambda \notin C_\alpha(0)$.

Remark 2.2 For a regular matrix $A = (a_{nk})$, note that

$$\overline{\lim_{n\to\infty}} \left(\sum_{k=0}^{\infty} |a_{nk}| \right) \leq 1$$

is equivalent to

$$\overline{\lim_{n\to\infty}} \left(\sum_{k=0}^{\infty} |a_{nk}| \right) = 1.$$

Remark 2.3 The proof of Theorem 2.1, for the case $\alpha = 1$, yields a very simple and elegant proof of Knopp's core theorem. This proof is much simpler than the proofs of Knopp's core theorem known earlier (for instance, see [3, p.149]).

2.3 Some Results for the Matrix Class (ℓ, ℓ)

We recall that

$$\ell = \left\{ x = \{x_k\} : \sum_{k=0}^{\infty} |x_k| < \infty \right\}.$$

Note that ℓ is a linear space with respect to coordinatewise addition and scalar multiplication and it is a Banach space with respect to the norm defined by

$$\|x\| = \sum_{k=0}^{\infty} |x_k|, \quad x = \{x_k\} \in \ell.$$

$(\ell, \ell; P)$ denotes the set of all infinite matrices $A = (a_{nk}) \in (\ell, \ell)$ such that

$$\sum_{n=0}^{\infty} (Ax)_n = \sum_{k=0}^{\infty} x_k, \quad x = \{x_k\} \in \ell.$$

We recall the following results (see [4–6])

Theorem 2.2 $A = (a_{nk}) \in (\ell, \ell)$ *if and only if*

$$\sup_{k \geq 0} \left(\sum_{n=0}^{\infty} |a_{nk}| \right) < \infty. \tag{2.5}$$

Further, $A \in (\ell, \ell; P)$ *if and only if* $A \in (\ell, \ell)$ *and (1.29) holds.*

Theorem 2.3 *The matrix class* (ℓ, ℓ) *is a Banach algebra under the norm*

$$\|A\| = \sup_{k \geq 0} \left(\sum_{n=0}^{\infty} |a_{nk}| \right), \quad A = (a_{nk}) \in (\ell, \ell), \tag{2.6}$$

with the usual matrix addition, scalar multiplication, and multiplication.

We shall now prove a few results for the matrix class (ℓ, ℓ) (see [7]).

Theorem 2.4 *The class* $(\ell, \ell; P)$*, as a subset of* (ℓ, ℓ)*, is a closed convex semigroup with identity, the multiplication being the usual matrix multiplication.*

Proof Let $A = (a_{nk}), B = (b_{nk}) \in (\ell, \ell; P)$ and $\lambda + \mu = 1$, λ, μ being nonnegative real numbers. Then, there exists $M > 0$ such that

$$\sup_{k \geq 0} \left(\sum_{n=0}^{\infty} |a_{nk}| \right), \sup_{k \geq 0} \left(\sum_{n=0}^{\infty} |b_{nk}| \right) \leq M.$$

Now,

$$\begin{aligned}\sup_{k\geq 0}\sum_{n=0}^{\infty}|\lambda a_{nk}+\mu b_{nk}| &\leq \lambda \sup_{k\geq 0}\left(\sum_{n=0}^{\infty}|a_{nk}|\right)+\mu \sup_{k\geq 0}\left(\sum_{n=0}^{\infty}|b_{nk}|\right)\\ &\leq (\lambda+\mu)M\\ &= M, \text{ since } \lambda+\mu=1.\end{aligned}$$

Also,

$$\begin{aligned}\sum_{n=0}^{\infty}(\lambda a_{nk}+\mu b_{nk}) &= \lambda\left(\sum_{n=0}^{\infty}a_{nk}\right)+\mu\left(\sum_{n=0}^{\infty}b_{nk}\right)\\ &= \lambda(1)+\mu(1)\\ &= \lambda+\mu\\ &= 1,\ k=0,1,2,\ldots,\end{aligned}$$

since $\sum_{n=0}^{\infty} a_{nk} = \sum_{n=0}^{\infty} b_{nk} = 1, k = 0, 1, 2, \ldots$, using (1.29). In view of Theorem 2.2, $\lambda A + \mu B \in (\ell, \ell; P)$ so that $(\ell, \ell; P)$ is a convex subset of (ℓ, ℓ).

Let, now, $A = (a_{nk}) \in \overline{(\ell, \ell; P)}$. Then, there exist $A^{(m)} = (a_{nk}^{(m)}), m = 0, 1, 2, \ldots$ such that

$$\|A^{(m)} - A\| \to 0, m \to \infty.$$

Thus, given $\epsilon > 0$, there exists a positive integer N such that

$$\|A^{(m)} - A\| < \infty, m \geq N,$$

$$i.e., \sup_{k\geq 0}\left(\sum_{n=0}^{\infty}|a_{nk}^{(m)} - a_{nk}|\right) < \epsilon, m \geq N. \tag{2.7}$$

Now,

$$\begin{aligned}\sup_{k\geq 0}\left(\sum_{n=0}^{\infty}|a_{nk}|\right) &\leq \sup_{k\geq 0}\left(\sum_{n=0}^{\infty}|a_{nk}-a_{nk}^{(N)}|\right)+\sup_{k\geq 0}\left(\sum_{n=0}^{\infty}|a_{nk}^{(N)}|\right)\\ &< \epsilon+\sup_{k\geq 0}\left(\sum_{n=0}^{\infty}|a_{nk}^{(N)}|\right), \text{ using (2.7)}\\ &< \infty,\end{aligned}$$

since $A^{(N)} \in (\ell, \ell; P)$, so that $A \in (\ell, \ell)$. Again,

$$
\begin{aligned}
\left|\sum_{n=0}^{\infty} a_{nk} - 1\right| &= \left|\sum_{n=0}^{\infty} a_{nk} - \sum_{n=0}^{\infty} a_{nk}^{(N)}\right|, \text{ since } A^{(N)} \in (\ell, \ell; P) \\
&= \left|\sum_{n=0}^{\infty} (a_{nk} - a_{nk}^{(N)})\right| \\
&\leq \sum_{n=0}^{\infty} |a_{nk} - a_{nk}^{(N)}| \\
&\leq \sup_{k \geq 0} \left(\sum_{n=0}^{\infty} |a_{nk} - a_{nk}^{(N)}|\right) \\
&< \epsilon, \ k = 0, 1, 2, \ldots, \text{ in view of (2.7)}
\end{aligned}
$$

so that

$$\left|\sum_{n=0}^{\infty} a_{nk} - 1\right| < \epsilon \text{ for all } \epsilon > 0.$$

Consequently,

$$\sum_{n=0}^{\infty} a_{nk} = 1, \ k = 0, 1, 2, \ldots.$$

Thus, $A \in (\ell, \ell; P)$ and so $(\ell, \ell; P)$ is a closed subset of (ℓ, ℓ).

It is clear that the unit matrix is in $(\ell, \ell; P)$, and it is the identity element of $(\ell, \ell; P)$.

To complete the proof, it suffices to check closure under matrix multiplication. If $A = (a_{nk})$, $B = (b_{nk}) \in (\ell, \ell; P)$, using Theorem 2.3, $AB \in (\ell, \ell)$. In fact, $AB \in (\ell, \ell; P)$, since

$$
\begin{aligned}
\sum_{n=0}^{\infty} c_{nk} &= \sum_{n=0}^{\infty} \left(\sum_{i=0}^{\infty} a_{ni} b_{ik}\right) \\
&= \sum_{i=0}^{\infty} b_{ik} \left(\sum_{n=0}^{\infty} a_{ni}\right) \\
&= \sum_{i=0}^{\infty} b_{ik}, \text{ since } \sum_{n=0}^{\infty} a_{ni} = 1, i = 0, 1, 2, \ldots \\
&= 1, \text{ since } \sum_{i=0}^{\infty} b_{ik} = 1, k = 0, 1, 2, \ldots.
\end{aligned}
$$

This completes the proof of the theorem. □

Remark 2.4 $(\ell, \ell; P)$ is not an algebra since the sum of two elements of $(\ell, \ell; P)$ is not in $(\ell, \ell; P)$.

We now introduce a convolution product (see [5]).

Definition 2.2 For $A = (a_{nk})$, $B = (b_{nk})$, define

$$(A * B)_{nk} = \sum_{i=0}^{n} a_{ik} b_{n-i,k}, \quad n, k = 0, 1, 2, \ldots. \tag{2.8}$$

$A * B = ((A * B)_{nk})$ is called the convolution product of A and B.

We keep the usual norm structure in (ℓ, ℓ) as defined by (2.6) and replace matrix product by the convolution product as defined by (2.8) and prove the following result.

Theorem 2.5 *(ℓ, ℓ) is a commutative Banach algebra, with identity, under the convolution product * as defined by (2.8). Furthermore, $(\ell, \ell; P)$, as a subset of (ℓ, ℓ), is a closed convex semigroup with identity.*

Proof Recall that it was proved in Theorem 2.4 that $(\ell, \ell; P)$ is a convex subset of (ℓ, ℓ). We will first prove closure under the convolution product *. Let $A = (a_{nk})$, $B = (b_{nk}) \in (\ell, \ell)$, and $A * B = (c_{nk})$. Then,

$$\begin{aligned}
\sum_{n=0}^{\infty} |c_{nk}| &= \sum_{n=0}^{\infty} \left| \sum_{i=0}^{n} a_{ik} b_{n-i,k} \right| \\
&\leq \sum_{n=0}^{\infty} \sum_{i=0}^{n} |a_{ik}| |b_{n-i,k}| \\
&= \left(\sum_{n=0}^{\infty} |a_{nk}| \right) \left(\sum_{n=0}^{\infty} |b_{nk}| \right) \\
&\leq \sup_{k \geq 0} \left(\sum_{n=0}^{\infty} |a_{nk}| \right) \sup_{k \geq 0} \left(\sum_{n=0}^{\infty} |b_{nk}| \right) \\
&= \|A\| \, \|B\|, \; k = 0, 1, 2, \ldots,
\end{aligned}$$

so that

$$\sup_{k \geq 0} \left(\sum_{n=0}^{\infty} |c_{nk}| \right) < \infty$$

and so $A * B \in (\ell, \ell)$. Also,

$$\|A * B\| \leq \|A\| \, \|B\|.$$

It is clear that $A * B = B * A$. The identity element is the matrix $E = (e_{nk})$, whose first row consists of 1's and which has 0's elsewhere,

$$i.e., \; e_{0k} = 1, \; k = 0, 1, 2, \ldots;$$
$$e_{nk} = 0, \; n = 1, 2, \ldots; k = 0, 1, 2, \ldots.$$

We note that $E \in (\ell, \ell; P)$ and $\|E\| = 1$. It now suffices to prove that $(\ell, \ell; P)$ is closed under the convolution product *. Now,

$$\begin{aligned}
\sum_{n=0}^{\infty} c_{nk} &= \sum_{n=0}^{\infty}\left(\sum_{i=0}^{n} a_{ik} b_{n-i,k}\right) \\
&= \left(\sum_{n=0}^{\infty} a_{nk}\right)\left(\sum_{n=0}^{\infty} b_{nk}\right) \\
&= 1, \; k = 0, 1, 2, \ldots,
\end{aligned}$$

where $A, B \in (\ell, \ell; P)$. This completes the proof of the theorem. □

2.4 A Mercerian Theorem

We close the present chapter by proving a Mercerian theorem for the Banach algebra (ℓ, ℓ) under the convolution product *.

Theorem 2.6 *If*

$$y_n = x_n + \lambda(c^n x_0 + c^{n-1} x_1 + \cdots + c x_{n-1} + x_n),$$

$|c| < 1$ *and if* $\{y_n\} \in \ell$, *then* $\{x_n\} \in \ell$, *provided*

$$|\lambda| < 1 - c.$$

Proof Since (ℓ, ℓ) is a Banach algebra under the convolution product *, if $|\lambda| < \frac{1}{\|A\|}$, $A \in (\ell, \ell)$, then $E - \lambda A$, where E is the identity element of (ℓ, ℓ) under *, has an inverse in (ℓ, ℓ). We recall that

$$E = (e_{nk}) = \begin{pmatrix} 1 & 1 & 1 & \cdots \\ 0 & 0 & 0 & \cdots \\ 0 & 0 & 0 & \cdots \\ \cdots & \cdots & \cdots & \cdots \end{pmatrix}.$$

We note that the equations

$$y_n = x_n + \lambda(c^n x_0 + c^{n-1} x_1 + \cdots + c x_{n-1} + x_n), \; |c| < 1, n = 0, 1, 2, \ldots$$

can be written in the form

$$(E + \lambda A) * x' = y',$$

where

$$A = \begin{pmatrix} 1 & 0 & 0 & \cdots \\ c & 0 & 0 & \cdots \\ c^2 & 0 & 0 & \cdots \\ \cdots & \cdots & \cdots & \cdots \end{pmatrix},$$

$$x' = \begin{pmatrix} x_0 & 0 & 0 & \cdots \\ x_1 & 0 & 0 & \cdots \\ x_2 & 0 & 0 & \cdots \\ \cdots & \cdots & \cdots & \cdots \end{pmatrix},$$

$$y' = \begin{pmatrix} y_0 & 0 & 0 & \cdots \\ y_1 & 0 & 0 & \cdots \\ y_2 & 0 & 0 & \cdots \\ \cdots & \cdots & \cdots & \cdots \end{pmatrix}.$$

It is clear that $A \in (\ell, \ell)$ with $\|A\| = \frac{1}{1-c}$. So, if $|\lambda| < 1 - c$, $(E + \lambda A)$ has an inverse in (ℓ, ℓ). Consequently, it follows that

$$x' = (E + \lambda A)^{-1} * y'.$$

Since $y' \in (\ell, \ell)$ and $(E + \lambda A)^{-1} \in (\ell, \ell)$, we have, $x' \in (\ell, \ell)$. In view of Theorem 2.2, it follows that $\{x_n\} \in \ell$, completing the proof of the theorem. □

References

1. Sherbakoff, A.A.: On cores of complex sequences and their regular transform. Mat. Zametki **22**, 815–828 (1977) (Russian)
2. Natarajan, P.N.: On the core of a sequence over valued fields. J. Indian. Math. Soc. **55**, 189–198 (1990)
3. Cooke, R.G.: Infinite Matrices and Sequence Spaces. Macmillan, New York (1950)
4. Fridy, J.A.: A note on absolute summability. Proc. Amer. Math. Soc. **20**, 285–286 (1969)
5. Knopp, K., Lorentz, G.G.: Beiträge zur absoluten Limitierung. Arch. Math. **2**, 10–16 (1949)
6. Maddox, I.J.: Elements of Functional Analysis. Cambridge University Press, Cambridge (1977)
7. Natarajan, P.N.: On the algebra (ℓ_1, ℓ_1) of infinite matrices. Analysis (München) **20**, 353–357 (2000)

Chapter 3
Special Summability Methods

In the current chapter, some special summability methods are introduced and their properties are studied. In the first section, we introduce the Weighted Mean method and study some of its properties. We prove a result, which gives an equivalent formulation of summability by Weighted Mean methods. The result of Hardy [3] and that of Móricz and Rhoades [4] are particular cases of the above result. The second section is devoted to a detailed study of the (M, λ_n) method or the Natarajan method. The connection between the Abel and the Natarajan methods is brought out in the third section. In the final section, the Euler method is introduced and its properties are studied. We prove an interesting product theorem involving the Euler and Natarajan methods.

3.1 Weighted Mean Method

The Weighted Mean method is defined as follows.

Definition 3.1 ([1, p. 16]) The Weighted Mean method or $(\overline{N}, p_n)$ method is defined by the infinite matrix $A = (a_{nk})$, where

$$a_{nk} = \begin{cases} \frac{p_k}{P_n}, & k \leq n; \\ 0, & k > n, \end{cases}$$

$P_n = \sum_{k=0}^{n} p_k, n = 0, 1, 2, \ldots, P_n \neq 0, n = 0, 1, 2, \ldots.$

Theorem 3.1 ([1, p. 16]) *The Weighted Mean method $(\overline{N}, p_n)$ is regular if and only if*

$$\sum_{k=0}^{n} |p_k| = O(P_n), \quad n \to \infty; \tag{3.1}$$

P.N. Natarajan, *Classical Summability Theory*, DOI 10.1007/978-981-10-4205-8_3

and

$$P_n \to \infty, \quad n \to \infty. \tag{3.2}$$

Remark 3.1

$$\begin{aligned}|P_n| &\leq \sum_{k=0}^{n} |p_k| \\ &\leq \sum_{k=0}^{n+m} |p_k| \\ &\leq L|P_{n+m}|,\end{aligned}$$

for some $L > 0$, $m = 0, 1, 2, \ldots; n = 0, 1, 2, \ldots$.

We now prove a result, which gives an equivalent formulation of summability by Weighted Mean methods.

Theorem 3.2 (see [2]) *Let $(\overline{N}, p_n)$, $(\overline{N}, q_n)$ be two regular Weighted Mean methods. For a given series $\sum_{k=0}^{\infty} x_k$, let*

$$b_n = q_n \sum_{k=n}^{\infty} \frac{x_k}{Q_k}, \quad n = 0, 1, 2, \ldots.$$

Let $\sum_{n=0}^{\infty} b_n$ converge to s. Then $\sum_{k=0}^{\infty} x_k$ is $(\overline{N}, p_n)$ summable to s if and only if

$$\sup_{n \geq 0} \left[\frac{1}{|P_n|} \sum_{k=1}^{n} \left| \frac{p_k Q_{k+1}}{q_{k+1}} - \frac{p_{k-1} Q_{k-1}}{q_k} \right| \right] < \infty. \tag{3.3}$$

Proof Let $\sum_{n=0}^{\infty} b_n$ converge to s. Then $B_n = \sum_{k=0}^{n} b_k \to s$, $n \to \infty$.

Now,

$$\begin{aligned}\frac{b_n}{q_n} - \frac{b_{n+1}}{q_{n+1}} &= \sum_{k=n}^{\infty} \frac{x_k}{Q_k} - \sum_{k=n+1}^{\infty} \frac{x_k}{Q_k} \\ &= \frac{x_n}{Q_n},\end{aligned}$$

so that

$$x_n = Q_n \left(\frac{b_n}{q_n} - \frac{b_{n+1}}{q_{n+1}} \right), \quad n = 0, 1, 2, \ldots.$$

Consequently,

$$\begin{aligned}
s_m &= \sum_{k=0}^{m} x_k \\
&= \sum_{k=0}^{m} Q_k \left(\frac{b_k}{q_k} - \frac{b_{k+1}}{q_{k+1}} \right) \\
&= \sum_{k=0}^{m} Q_k \frac{b_k}{q_k} - \sum_{k=1}^{m+1} Q_{k-1} \frac{b_k}{q_k} \\
&= Q_0 \frac{b_0}{q_0} + \sum_{k=1}^{m} (Q_k - Q_{k-1}) \frac{b_k}{q_k} - Q_m \frac{b_{m+1}}{q_{m+1}} \\
&= b_0 + \sum_{k=1}^{m} q_k \frac{b_k}{q_k} - Q_m \frac{b_{m+1}}{q_{m+1}} \\
&= b_0 + \sum_{k=1}^{m} b_k - Q_m \frac{b_{m+1}}{q_{m+1}} \\
&= \sum_{k=0}^{m} b_k - Q_m \frac{b_{m+1}}{q_{m+1}} \\
&= B_m - Q_m \frac{b_{m+1}}{q_{m+1}}. \qquad (3.4)
\end{aligned}$$

By hypothesis, $\sum_{k=0}^{\infty} \frac{x_k}{Q_k}$ converges so that

$$\frac{b_n}{q_n} = \sum_{k=n}^{\infty} \frac{x_k}{Q_k} \to 0, \quad n \to \infty.$$

Now,

$$\frac{s_m}{Q_m} = \frac{B_m}{Q_m} - \frac{b_{m+1}}{q_{m+1}}, \quad \text{using (3.4)}.$$

Since $\{B_n\}$ converges, it is bounded so that $|B_n| \leq M, n = 0, 1, 2, \ldots$, for some $M > 0$. Since $(\overline{N}, q_n)$ is regular, $|Q_n| \to \infty, n \to \infty$ so that

$$\left| \frac{B_m}{Q_m} \right| \leq \frac{M}{|Q_m|} \to 0, \quad m \to \infty.$$

Thus,

$$\frac{s_m}{Q_m} \to 0, \quad m \to \infty.$$

Now, for $n = 0, 1, 2, \ldots,$

$$
\begin{aligned}
b_n &= q_n \sum_{k=n}^{\infty} \frac{x_k}{Q_k} \\
&= q_n \lim_{m\to\infty} \sum_{k=n}^{m} \frac{x_k}{Q_k} \\
&= q_n \lim_{m\to\infty} \sum_{k=n}^{m} \frac{s_k - s_{k-1}}{Q_k}, \quad s_{-1} = 0 \\
&= q_n \lim_{m\to\infty} \left\{ \sum_{k=n}^{m} \frac{s_k}{Q_k} - \sum_{k=n-1}^{m-1} \frac{s_k}{Q_{k+1}} \right\} \\
&= q_n \lim_{m\to\infty} \left\{ \sum_{k=n}^{m-1} \frac{s_k}{Q_k} + \frac{s_m}{Q_m} - \sum_{k=n}^{m-1} \frac{s_k}{Q_{k+1}} - \frac{s_{n-1}}{Q_n} \right\} \\
&= q_n \lim_{m\to\infty} \left\{ \sum_{k=n}^{m-1} \left(\frac{1}{Q_k} - \frac{1}{Q_{k+1}} \right) s_k + \frac{s_m}{Q_m} - \frac{s_{n-1}}{Q_n} \right\} \\
&= -q_n \frac{s_{n-1}}{Q_n} + q_n \sum_{k=n}^{\infty} \left(\frac{1}{Q_k} - \frac{1}{Q_{k+1}} \right) s_k, \text{ since } \lim_{m\to\infty} \frac{s_m}{Q_m} = 0 \\
&= -q_n \frac{s_{n-1}}{Q_n} + q_n \sum_{k=n}^{\infty} c_k s_k, \text{ where } c_k = \frac{1}{Q_k} - \frac{1}{Q_{k+1}}, k = 0, 1, 2, \ldots. \qquad (3.5)
\end{aligned}
$$

We now have,

$$
\begin{aligned}
B_n &= \sum_{k=0}^{n-1} b_k + b_n \\
&= \sum_{k=0}^{n-1} \frac{b_k}{q_k} q_k + b_n \\
&= \sum_{k=0}^{n-1} q_k \left(\sum_{u=k}^{\infty} \frac{x_u}{Q_u} \right) + b_n \\
&= q_0 \sum_{u=0}^{\infty} \frac{x_u}{Q_u} + q_1 \sum_{u=1}^{\infty} \frac{x_u}{Q_u} + q_2 \sum_{u=2}^{\infty} \frac{x_u}{Q_u} + \cdots + q_{n-1} \sum_{u=n-1}^{\infty} \frac{x_u}{Q_u} + b_n \\
&= (q_0 + q_1 + \cdots + q_{n-1}) \sum_{u=n-1}^{\infty} \frac{x_u}{Q_u} + q_0 \sum_{u=0}^{n-2} \frac{x_u}{Q_u} + q_1 \sum_{u=1}^{n-2} \frac{x_u}{Q_u} + q_2 \sum_{u=2}^{n-2} \frac{x_u}{Q_u} \\
&\quad + \cdots + q_{n-2} \frac{x_{n-2}}{Q_{n-2}} + b_n
\end{aligned}
$$

$$
\begin{aligned}
&= Q_{n-1} \sum_{u=n-1}^{\infty} \frac{x_u}{Q_u} + b_n + \frac{x_{n-2}}{Q_{n-2}} Q_{n-2} + \frac{x_{n-3}}{Q_{n-3}} Q_{n-3} + \cdots + \frac{x_0}{Q_0} Q_0 \\
&= Q_{n-1} \sum_{u=n-1}^{\infty} \frac{x_u}{Q_u} + b_n + \sum_{k=0}^{n-2} x_k \\
&= s_{n-2} + b_n + Q_{n-1} \sum_{u=n-1}^{\infty} \frac{x_u}{Q_u} \\
&= s_{n-2} + q_n \sum_{u=n}^{\infty} \frac{x_u}{Q_u} + Q_{n-1} \sum_{u=n-1}^{\infty} \frac{x_u}{Q_u} \\
&= s_{n-2} + (Q_n - Q_{n-1}) \sum_{u=n}^{\infty} \frac{x_u}{Q_u} + Q_{n-1} \sum_{u=n-1}^{\infty} \frac{x_u}{Q_u} \\
&= s_{n-2} + Q_n \sum_{u=n}^{\infty} \frac{x_u}{Q_u} + Q_{n-1} \frac{x_{n-1}}{Q_{n-1}} \\
&= s_{n-2} + Q_n \sum_{u=n}^{\infty} \frac{x_u}{Q_u} + x_{n-1} \\
&= s_{n-1} + Q_n \sum_{u=n}^{\infty} \frac{x_u}{Q_u} \\
&= s_{n-1} + Q_n \frac{b_n}{q_n} \\
&= s_{n-1} + Q_n \left[-\frac{s_{n-1}}{Q_n} + \sum_{k=n}^{\infty} c_k s_k \right], \text{ using (3.5)} \\
&= s_{n-1} - s_{n-1} + Q_n \sum_{k=n}^{\infty} c_k s_k \\
&= Q_n \sum_{k=n}^{\infty} c_k s_k,
\end{aligned}
$$

so that

$$
\frac{B_n}{Q_n} = \sum_{k=n}^{\infty} c_k s_k .
$$

Consequently,

$$
c_n s_n = \frac{B_n}{Q_n} - \frac{B_{n+1}}{Q_{n+1}}, \quad n = 0, 1, 2, \ldots . \tag{3.6}
$$

If $\{T_n\}$ is the $(\overline{N}, p_n)$ transform of $\{s_k\}$, then

$$
\begin{aligned}
T_n &= \frac{1}{P_n}\sum_{k=0}^{n} p_k s_k \\
&= \frac{1}{P_n}\sum_{k=0}^{n} p_k \frac{1}{c_k}\left\{\frac{B_k}{Q_k}-\frac{B_{k+1}}{Q_{k+1}}\right\}, \text{ using (3.6)} \\
&= \frac{1}{P_n}\left[\frac{p_0}{c_0}\frac{B_0}{Q_0}+\sum_{k=1}^{n}\left\{\frac{p_k}{c_k}-\frac{p_{k-1}}{c_{k-1}}\right\}\frac{B_k}{Q_k}-\frac{p_n}{c_n}\frac{B_{n+1}}{Q_{n+1}}\right] \\
&= \sum_{k=0}^{\infty} a_{nk} B_k,
\end{aligned}
$$

where

$$
a_{nk} = \begin{cases} \frac{1}{P_n}\frac{p_0}{c_0 Q_0}, & k=0; \\ \frac{1}{P_n}\left\{\frac{p_k}{c_k}-\frac{p_{k-1}}{c_{k-1}}\right\}\frac{1}{Q_k}, & 1 \le k \le n; \\ -\frac{1}{P_n}\frac{p_n}{c_n Q_{n+1}}, & k=n+1; \\ 0, & k \ge n+2. \end{cases}
$$

Note that $\lim_{n\to\infty} a_{nk} = 0, k = 0, 1, 2, \ldots$. Also,

$$
\begin{aligned}
\sum_{k=0}^{\infty} a_{nk} &= \sum_{k=0}^{n+1} a_{nk} \\
&= \frac{1}{P_n}\left[\frac{p_0}{c_0 Q_0}+\sum_{k=1}^{n}\left\{\frac{p_k}{c_k}-\frac{p_{k-1}}{c_{k-1}}\right\}\frac{1}{Q_k}-\frac{p_n}{c_n Q_{n+1}}\right] \\
&= \frac{1}{P_n}\left[\frac{p_0}{c_0 Q_0}+\left(\frac{p_1}{c_1}-\frac{p_0}{c_0}\right)\frac{1}{Q_1}+\left(\frac{p_2}{c_2}-\frac{p_1}{c_1}\right)\frac{1}{Q_2}\right. \\
&\qquad \left. +\cdots+\left(\frac{p_n}{c_n}-\frac{p_{n-1}}{c_{n-1}}\right)\frac{1}{Q_n}-\frac{p_n}{c_n Q_{n+1}}\right] \\
&= \frac{1}{P_n}\left[\frac{p_0}{c_0}\left(\frac{1}{Q_0}-\frac{1}{Q_1}\right)+\frac{p_1}{c_1}\left(\frac{1}{Q_1}-\frac{1}{Q_2}\right)+\frac{p_2}{c_2}\left(\frac{1}{Q_2}-\frac{1}{Q_3}\right)\right. \\
&\qquad \left. +\cdots+\frac{p_n}{c_n}\left(\frac{1}{Q_n}-\frac{1}{Q_{n+1}}\right)\right] \\
&= \frac{1}{P_n}\left[\frac{p_0}{c_0}c_0+\frac{p_1}{c_1}c_1+\cdots+\frac{p_n}{c_n}c_n\right] \\
&= \frac{1}{P_n}[p_0+p_1+\cdots+p_n] \\
&= \frac{1}{P_n}P_n \\
&= 1, \quad n = 0, 1, 2, \ldots,
\end{aligned}
$$

so that $\lim_{n\to\infty} \sum_{k=0}^{\infty} a_{nk} = 1$. By hypothesis, $B_k \to s$, $k \to \infty$. In view of Theorem 1.1, $T_n \to s$, $n \to \infty$, i.e., $\sum_{n=0}^{\infty} x_n$ is $(\overline{N}, p_n)$ summable to s if and only if

$$\sup_{n\geq 0} \frac{1}{|P_n|} \left[\left| \frac{p_0}{c_0 Q_0} \right| + \left| \frac{p_n}{c_n Q_{n+1}} \right| + \sum_{k=1}^{n} \left| \frac{1}{Q_k} \left\{ \frac{p_k}{c_k} - \frac{p_{k-1}}{c_{k-1}} \right\} \right| \right] < \infty. \tag{3.7}$$

In view of Remark 3.1, we have,

$$\begin{aligned} |P_n| &\leq L|P_{n+m}|, \\ |Q_n| &\leq L|Q_{n+m}|, \end{aligned} \tag{3.8}$$

for some $L > 0$, $m = 0, 1, 2, \ldots; n = 0, 1, 2, \ldots$. However,

$$\begin{aligned} \left| \frac{p_n}{P_n c_n Q_{n+1}} \right| &\leq L \left| \frac{p_n}{P_n c_n Q_n} \right|, \text{ using (3.8)} \\ &= \frac{L}{|P_n Q_n|} \left| \frac{p_n}{c_n} \right| \\ &= \frac{L}{|P_n Q_n|} \left| \sum_{k=1}^{n} \left(\frac{p_k}{c_k} - \frac{p_{k-1}}{c_{k-1}} \right) + \frac{p_0}{c_0} \right| \\ &\leq \frac{L^2}{|P_n|} \left| \sum_{k=1}^{n} \frac{1}{Q_k} \left(\frac{p_k}{c_k} - \frac{p_{k-1}}{c_{k-1}} \right) + \frac{p_0}{c_0 Q_0} \right|, \\ &\qquad \text{since } |Q_k| \leq L|Q_n|, k \leq n, \text{ using (3.8) again} \\ &\leq \frac{L^2}{|P_n|} \left[\sum_{k=1}^{n} \left| \frac{1}{Q_k} \left(\frac{p_k}{c_k} - \frac{p_{k-1}}{c_{k-1}} \right) \right| + \left| \frac{p_0}{c_0 Q_0} \right| \right]. \end{aligned}$$

Thus, (3.7) is equivalent to

$$\sup_{n\geq 0} \frac{1}{|P_n|} \left[\sum_{k=1}^{n} \left| \frac{1}{Q_k} \left\{ \frac{p_k}{c_k} - \frac{p_{k-1}}{c_{k-1}} \right\} \right| \right] < \infty. \tag{3.9}$$

Now,

$$\begin{aligned} \frac{p_k}{c_k} - \frac{p_{k-1}}{c_{k-1}} &= \frac{p_k}{\frac{1}{Q_k} - \frac{1}{Q_{k+1}}} - \frac{p_{k-1}}{\frac{1}{Q_{k-1}} - \frac{1}{Q_k}} \\ &= \frac{p_k Q_k Q_{k+1}}{q_{k+1}} - \frac{p_{k-1} Q_k Q_{k-1}}{q_k} \end{aligned}$$

so that (3.9) can now be written as

$$\sup_{n\geq 0}\frac{1}{|P_n|}\left[\sum_{k=1}^{n}\left|\frac{1}{Q_k}\left\{\frac{p_kQ_kQ_{k+1}}{q_{k+1}}-\frac{p_{k-1}Q_kQ_{k-1}}{q_k}\right\}\right|\right]<\infty,$$

$$i.e., \sup_{n\geq 0}\frac{1}{|P_n|}\left[\sum_{k=1}^{n}\left|\frac{p_kQ_{k+1}}{q_{k+1}}-\frac{p_{k-1}Q_{k-1}}{q_k}\right|\right]<\infty,$$

completing the proof of the theorem. □

Remark 3.2 The result of Hardy [3] and that of Móricz and Rhoades [4] are particular cases of Theorem 3.2. In the context of Theorem 3.2, it is worthwhile to note that the result of Móricz and Rhoades is valid even without the assumption $\frac{p_n}{P_n}\to 0, n\to\infty$.

In the context of Theorem 3.2, we have another interesting result.

Theorem 3.3 (see [5]) *Let $(\overline{N}, p_n)$, $(\overline{N}, q_n)$ be two regular Weighted Mean methods and*

$$P_n = O(p_nQ_n),\ \ n\to\infty, \tag{3.10}$$

$$i.e., \left|\frac{P_n}{p_nQ_n}\right|\leq M,\ \ n=0,1,2,\ldots,\ for\ some\ M>0.$$

Let $\sum_{n=0}^{\infty}x_n$ be $(\overline{N}, p_n)$ summable to s. Then $\sum_{n=0}^{\infty}b_n$ converges to s if and only if

$$\sup_{n\geq 0}\left[|Q_n|\sum_{k=n}^{\infty}\left|\frac{P_k}{Q_{k+1}}\left(\frac{q_k}{p_kQ_k}-\frac{q_{k+2}}{p_{k+1}Q_{k+2}}\right)\right|\right]<\infty, \tag{3.11}$$

where

$$b_n = q_n\sum_{k=n}^{\infty}\frac{x_k}{Q_k},\ \ n=0,1,2,\ldots.$$

Proof Let

$$s_n = \sum_{k=0}^{n}x_k,$$

$$t_n = \frac{p_0s_0+p_1s_1+\cdots+p_ns_n}{P_n},\ \ n=0,1,2,\ldots.$$

Then,

$$
\begin{aligned}
s_0 &= t_0,\\
s_n &= \frac{1}{p_n}(P_n t_n - P_{n-1}t_{n-1}), \quad n = 1, 2, \ldots.
\end{aligned}
$$

Let $\sum_{n=0}^{\infty} x_n$ be $(\overline{N}, p_n)$ summable to s, i.e., $\lim_{n\to\infty} t_n = s$.

Now,

$$
\begin{aligned}
\frac{s_n}{Q_n} &= \frac{1}{p_n Q_n}(P_n t_n - P_{n-1}t_{n-1})\\
&= \frac{1}{p_n Q_n}[P_n(t_n - s) - P_{n-1}(t_{n-1} - s) + s(P_n - P_{n-1})]\\
&= \frac{1}{p_n Q_n}[P_n(t_n - s) - P_{n-1}(t_{n-1} - s) + s p_n]\\
&= \frac{P_n}{p_n Q_n}(t_n - s) - \frac{P_{n-1}}{p_n Q_n}(t_{n-1} - s) + \frac{s}{Q_n},
\end{aligned}
$$

so that

$$
\begin{aligned}
\left|\frac{s_n}{Q_n}\right| &\le M[|t_n - s| + L|t_{n-1} - s|] + \frac{|s|}{|Q_n|},\\
&\quad \text{since } |P_{n-1}| \le L|P_n|, \text{ using Remark 3.1}\\
&\to 0, n \to \infty, \text{ since } \lim_{n\to\infty} t_n = s \text{ and } \lim_{n\to\infty} |Q_n| = \infty,\\
&\quad (\overline{N}, q_n) \text{ being regular, using Theorem 3.1.}
\end{aligned}
$$

Now, in view of (3.5),

$$
b_n = -\frac{q_n s_{n-1}}{Q_n} + q_n \sum_{k=n}^{\infty} c_k s_k,
$$

where

$$
c_k = \frac{1}{Q_k} - \frac{1}{Q_{k+1}}, \quad k = 0, 1, 2, \ldots.
$$

We now have,

$$
\begin{aligned}
B_n &= Q_n \sum_{k=n}^{\infty} c_k s_k \text{ as worked out in Theorem 3.2}\\
&= Q_n \lim_{m\to\infty} \sum_{k=n}^{m} c_k s_k
\end{aligned}
$$

$$= Q_n \lim_{m\to\infty} \sum_{k=n}^{m} c_k \frac{1}{p_k}\{P_k t_k - P_{k-1}t_{k-1}\}$$

$$= Q_n \lim_{m\to\infty} \left[\frac{c_m P_m t_m}{p_m} - \frac{c_n P_{n-1} t_{n-1}}{p_n} + \sum_{k=n}^{m-1} P_k t_k \left(\frac{c_k}{p_k} - \frac{c_{k+1}}{p_{k+1}} \right) \right]. \tag{3.12}$$

Let

$$A_1 = \left\{ \{x_k\} : \sum_{k=0}^{\infty} x_k \text{ is } (\overline{N}, p_n) \text{ summable} \right\};$$

$$A_2 = \left\{ \{x_k\} : \sum_{k=0}^{\infty} b_k \text{ converges} \right\}.$$

We note that A_1, A_2 are BK spaces with respect to the norms defined by

$$\|x\|_{A_1} = \sup_{n\geq 0} |t_n|, \ x = \{x_k\} \in A_1;$$

and

$$\|x\|_{A_2} = \sup_{n\geq 0} |B_n|, \ x = \{x_k\} \in A_2,$$

respectively. Appealing to Banach–Steinhaus theorem (see [6]), we have,

$$\|x\|_{A_2} \leq U \|x\|_{A_1}, \quad \text{for some } U > 0. \tag{3.13}$$

For every fixed $k = 0, 1, 2, \ldots$, define the sequence $x = \{x_k\}$, where

$$x_n = \begin{cases} 1, & \text{if } n = k; \\ -1, & \text{if } n = k+1; \\ 0, & \text{otherwise.} \end{cases}$$

For this sequence x,

$$\|x\|_{A_1} = \left| \frac{p_k}{P_k} \right| \text{ and } \|x\|_{A_2} = |Q_k c_k|.$$

Using (3.13), we have, for $k = 0, 1, 2, \ldots$,

$$|Q_k c_k| \leq U \left| \frac{p_k}{P_k} \right|,$$

so that

$$\left|\frac{c_k P_k}{p_k}\right| \leq \frac{U}{|Q_k|}$$
$$\to 0,\ k \to \infty,\ \text{since } \lim_{k\to\infty} |Q_k| = \infty,\ \text{in view of Theorem 3.1.}$$

Thus,

$$\lim_{k\to\infty} \frac{c_k P_k}{p_k} = 0. \tag{3.14}$$

Using (3.14) in (3.12), we have,

$$B_n = -\frac{c_n P_{n-1} t_{n-1}}{p_n} Q_n + Q_n \sum_{k=n}^{\infty} P_k t_k \left(\frac{c_k}{p_k} - \frac{c_{k+1}}{p_{k+1}}\right)$$
$$= \sum_{k=n}^{\infty} a_{nk} t_k,$$

where (a_{nk}) is defined by

$$a_{nk} = \begin{cases} 0, & \text{if } 0 \leq k < n-1; \\ -\frac{Q_n c_n P_{n-1}}{p_n}, & \text{if } k = n-1; \\ Q_n P_k \left(\frac{c_k}{p_k} - \frac{c_{k+1}}{p_{k+1}}\right), & \text{if } k \geq n. \end{cases}$$

First we note that $\lim_{n\to\infty} a_{nk} = 0$, $k = 0, 1, 2, \ldots$. We also note that

$$\sum_{k=0}^{\infty} a_{nk} = 1, \quad n = 0, 1, 2, \ldots,$$

so that

$$\lim_{n\to\infty} \sum_{k=0}^{\infty} a_{nk} = 1.$$

Thus, appealing to Theorem 1.1, $\sum_{n=0}^{\infty} b_n$ converges to s if and only if

$$\sup_{n\geq 0} \left[|Q_n| \left\{ \left|\frac{c_n P_{n-1}}{p_n}\right| + \sum_{k=n}^{\infty} \left| P_k \left(\frac{c_k}{p_k} - \frac{c_{k+1}}{p_{k+1}}\right)\right| \right\} \right] < \infty. \tag{3.15}$$

However,

$$\left|\frac{Q_n c_n P_{n-1}}{p_n}\right| \le L \left|\frac{Q_n c_n P_n}{p_n}\right|,$$

$$\text{since } |P_{n-1}| \le L|P_n|, \quad \text{using Remark 3.1}$$

$$= L|Q_n| \left|P_n \sum_{k=n}^{\infty} \left(\frac{c_k}{p_k} - \frac{c_{k+1}}{p_{k+1}}\right)\right|, \quad \text{using (3.14)}$$

$$\le L^2|Q_n| \left|\sum_{k=n}^{\infty} P_k \left(\frac{c_k}{p_k} - \frac{c_{k+1}}{p_{k+1}}\right)\right|,$$

$$\text{since } |P_n| \le L|P_k|, k \ge n, \quad \text{using Remark 3.1 again}$$

$$\le L^2|Q_n| \sum_{k=n}^{\infty} \left|P_k \left(\frac{c_k}{p_k} - \frac{c_{k+1}}{p_{k+1}}\right)\right|. \tag{3.16}$$

Using (3.16), it is clear that (3.15) is equivalent to

$$\sup_{n\ge 0} |Q_n| \left[\sum_{k=n}^{\infty} \left|P_k \left(\frac{c_k}{p_k} - \frac{c_{k+1}}{p_{k+1}}\right)\right|\right] < \infty.$$

Now,

$$\frac{c_k}{p_k} - \frac{c_{k+1}}{p_{k+1}} = \frac{1}{p_k}\left(\frac{1}{Q_k} - \frac{1}{Q_{k+1}}\right) - \frac{1}{p_{k+1}}\left(\frac{1}{Q_{k+1}} - \frac{1}{Q_{k+2}}\right)$$
$$= \frac{q_{k+1}}{p_k Q_k Q_{k+1}} - \frac{q_{k+2}}{p_{k+1} Q_{k+1} Q_{k+2}}.$$

So $\sum_{n=0}^{\infty} b_n$ converges if and only if

$$\sup_{n\ge 0} |Q_n| \left[\sum_{k=n}^{\infty} \left|P_k \left(\frac{q_{k+1}}{p_k Q_k Q_{k+1}} - \frac{q_{k+2}}{p_{k+1} Q_{k+1} Q_{k+2}}\right)\right|\right] < \infty,$$

$$\textit{i.e.}, \sup_{n\ge 0} |Q_n| \left[\sum_{k=n}^{\infty} \left|\frac{P_k}{Q_{k+1}} \left(\frac{q_{k+1}}{p_k Q_k} - \frac{q_{k+2}}{p_{k+1} Q_{k+2}}\right)\right|\right] < \infty,$$

completing the proof of the theorem. □

3.2 (M, λ_n) Method or Natarajan Method

Natarajan introduced the (M, λ_n) method and studied some of its properties in [7–9].

Definition 3.2 Let $\{\lambda_n\}$ be a sequence such that $\sum_{n=0}^{\infty} |\lambda_n| < \infty$. The (M, λ_n) method or the Natarajan method is defined by the infinite matrix (a_{nk}), where

$$a_{nk} = \begin{cases} \lambda_{n-k}, & k \leq n; \\ 0, & k > n. \end{cases}$$

Remark 3.3 In this context, we note that the (M, λ_n) method reduces to the well-known Y-method, when $\lambda_0 = \lambda_1 = \frac{1}{2}$ and $\lambda_n = 0$, $n \geq 2$.

The Natarajan method (M, λ_n) is a non-trivial summability method, i.e., it is not equivalent to convergence. Take any (M, λ_n) method. Then, we have, $\sum_{n=0}^{\infty} |\lambda_n| < \infty$. Consider the sequence $\{1, 0, 1, 0, \dots\}$, which is not convergent. If $\{\sigma_n\}$ is the (M, λ_n)-transform of $\{1, 0, 1, 0, \dots\}$, then,

$$\sigma_n = \lambda_0 + \lambda_2 + \lambda_4 + \cdots + \lambda_{2k}, \quad \text{if } n = 2k \text{ or } 2k + 1.$$

Now,

$$\sum_{k=0}^{\infty} |\lambda_{2k}| \leq \sum_{n=0}^{\infty} |\lambda_n| < \infty,$$

so that $\{\sigma_n\}$ converges to s (say). Thus, $\{1, 0, 1, 0, \dots\}$ is (M, λ_n)-summable to s. Similarly, the series $1-1+1-1+\cdots$, whose partial sum sequence is $\{1, 0, 1, 0, \dots\}$, is (M, λ_n)-summable. In particular, if $\lambda_0 = \lambda_1 = \frac{1}{2}$, $\lambda_n = 0$, $n \geq 2$, the (M, λ_n) method reduces to the Y-method. The non-convergent sequence $\{1, 0, 1, 0, \dots\}$ and the non-convergent series $1 - 1 + 1 - 1 + \cdots$ are Y-summable to $\frac{1}{2}$. The reader should give other examples of non-convergent sequences and non-convergent series which are (M, λ_n)-summable.

Theorem 3.4 *The (M, λ_n) method is regular if and only if*

$$\sum_{n=0}^{\infty} \lambda_n = 1.$$

Proof Since $\sum_{n=0}^{\infty} |\lambda_n| < \infty$, $\sup_{n\geq 0} \sum_{k=0}^{\infty} |a_{nk}| = \sup_{n\geq 0} \sum_{k=0}^{n} |a_{nk}| = \sup_{n\geq 0} \sum_{k=0}^{n} |\lambda_{n-k}| = \sup_{n\geq 0} \sum_{k=0}^{n} |\lambda_k| < \infty$ and $\lim_{n\to\infty} a_{nk} = \lim_{n\to\infty} \lambda_{n-k} = 0$, $k = 0, 1, 2, \dots$, since $\lim_{n\to\infty} \lambda_n = 0$. Thus, $(M, \lambda_n) \equiv (a_{nk})$ is regular if and only if

$$\begin{aligned}1 = \lim_{n\to\infty} \sum_{k=0}^{\infty} a_{nk} &= \lim_{n\to\infty} \sum_{k=0}^{n} a_{nk} \\ &= \lim_{n\to\infty} \sum_{k=0}^{n} \lambda_{n-k} \\ &= \lim_{n\to\infty} \sum_{k=0}^{n} \lambda_k \\ &= \sum_{k=0}^{\infty} \lambda_k,\end{aligned}$$

completing the proof. □

Definition 3.3 Two matrix methods $A = (a_{nk})$, $B = (b_{nk})$ are said to be consistent if whenever $x = \{x_k\}$ is A-summable to s and B-summable to t, then $s = t$.

We now have,

Theorem 3.5 *Any two regular methods* (M, λ_n), (M, μ_n) *are consistent.*

Proof Let (M, λ_n), (M, μ_n) be two regular methods. Let

$$u_n = \lambda_0 x_n + \lambda_1 x_{n-1} + \cdots + \lambda_n x_0 \to s, \quad n \to \infty$$

and

$$v_n = \mu_0 x_n + \mu_1 x_{n-1} + \cdots + \mu_n x_0 \to t, \quad n \to \infty.$$

Let

$$\gamma_n = \lambda_0 \mu_n + \lambda_1 \mu_{n-1} + \cdots + \lambda_n \mu_0, \quad n = 0, 1, 2, \ldots.$$

Now,

$$\begin{aligned}w_n &= \gamma_0 x_n + \gamma_1 x_{n-1} + \cdots + \gamma_n x_0 \\ &= (\lambda_0\mu_0)x_n + (\lambda_0\mu_1 + \lambda_1\mu_0)x_{n-1} + \cdots + (\lambda_0\mu_n + \lambda_1\mu_{n-1} + \cdots + \lambda_n\mu_0)x_0 \\ &= \lambda_0(\mu_0 x_n + \mu_1 x_{n-1} + \cdots + \mu_n x_0) + \lambda_1(\mu_0 x_{n-1} + \mu_1 x_{n-2} + \cdots + \mu_{n-1} x_0) \\ &\quad + \cdots + \lambda_n(\mu_0 x_0) \\ &= \lambda_0 v_n + \lambda_1 v_{n-1} + \cdots + \lambda_n v_0.\end{aligned}$$

Thus, $\{w_n\}$ is the (M, λ_n)-transform of the sequence $\{v_k\}$. Since (M, λ_n) is regular and $\lim_{k\to\infty} v_k = t$, it follows that

$$\lim_{n\to\infty} w_n = t.$$

In a similar manner, we can prove that

$$\lim_{n\to\infty} w_n = s,$$

so that $s = t$, completing the proof. □

Definition 3.4 Let $s = \{s_0, s_1, s_2, \dots\}, \overline{s} = \{0, s_0, s_1, s_2, \dots\}$ and $s^* = \{s_1, s_2, \dots\}$. The summability method A is said to be "translative" if $\overline{s}$, s^* are A-summable to t whenever s is A-summable to t.

Theorem 3.6 *Every (M, λ_n) method is translative.*

Proof Writing $A \equiv (M, \lambda_n)$,

$$\begin{aligned}(A\overline{s})_n &= \lambda_n 0 + \lambda_{n-1}s_0 + \lambda_{n-2}s_1 + \cdots + \lambda_0 s_{n-1} \\ &= \lambda_{n-1}s_0 + \lambda_{n-2}s_1 + \cdots + \lambda_0 s_{n-1} \\ &= u_{n-1},\end{aligned}$$

where

$$u_n = \lambda_n s_0 + \lambda_{n-1}s_1 + \cdots + \lambda_0 s_n, \ \ n = 0, 1, 2, \dots.$$

So, if $u_n \to t, n \to \infty$, then

$$(A\overline{s})_n \to t, \ \ n \to \infty.$$

Also,

$$\begin{aligned}(As^*)_n &= \lambda_n s_1 + \lambda_{n-1}s_2 + \cdots + \lambda_0 s_{n+1} \\ &= (\lambda_{n+1}s_0 + \lambda_n s_1 + \cdots + \lambda_0 s_{n+1}) - \lambda_{n+1}s_0 \\ &= u_{n+1} - \lambda_{n+1}s_0 \\ &\to t, \ \ n \to \infty,\end{aligned}$$

since $u_n \to t, n \to \infty$ and $\lambda_n \to 0, n \to \infty$. Thus, (M, λ_n) is translative. □

Definition 3.5 Given two summability methods A, B, we say that A is included in B (or B includes A), written as

$$A \subseteq B \text{ (or } B \supseteq A),$$

if whenever $x = \{x_k\}$ is A-summable to s, then it is also B-summable to s.

We now have

Theorem 3.7 (Inclusion theorem) *Given the methods (M, λ_n), (M, μ_n),*

$$(M, \lambda_n) \subseteq (M, \mu_n)$$

if and only if

$$\sum_{n=0}^{\infty} |k_n| < \infty \textit{ and } \sum_{n=0}^{\infty} k_n = 1,$$

$$\textit{where } \frac{\mu(x)}{\lambda(x)} = k(x) = \sum_{n=0}^{\infty} k_n x^n, \ \lambda(x) = \sum_{n=0}^{\infty} \lambda_n x^n, \ \mu(x) = \sum_{n=0}^{\infty} \mu_n x^n.$$

Proof As in Hardy [10, pp. 65–68], we can work out as follows:

Let $\lambda(x) = \sum_{n=0}^{\infty} \lambda_n x^n$, $\mu(x) = \sum_{n=0}^{\infty} \mu_n x^n$. Both the series on the right converge for $|x| < 1$. Let $\{u_n\}$, $\{v_n\}$ be the (M, λ_n), (M, μ_n) transforms of $\{s_n\}$, respectively. If $|x| < 1$, then

$$\begin{aligned}
\sum_{n=0}^{\infty} v_n x^n &= \sum_{n=0}^{\infty} (\mu_0 s_n + \mu_1 s_{n-1} + \cdots + \mu_n s_0) x^n \\
&= \left(\sum_{n=0}^{\infty} \mu_n x^n\right)\left(\sum_{n=0}^{\infty} s_n x^n\right) \\
&= \mu(x) s(x).
\end{aligned}$$

Similarly,

$$\sum_{n=0}^{\infty} u_n x^n = \lambda(x) s(x), \quad \text{if } |x| < 1.$$

Now,

$$\begin{aligned}
k(x)\lambda(x) &= \mu(x), \\
k(x)\lambda(x)s(x) &= \mu(x)s(x), \\
\textit{i.e., } k(x)\left(\sum_{n=0}^{\infty} u_n x^n\right) &= \sum_{n=0}^{\infty} v_n x^n.
\end{aligned}$$

Thus,

$$\begin{aligned}
v_n &= k_0 u_n + k_1 u_{n-1} + \cdots + k_n u_0 \\
&= \sum_{j=0}^{\infty} a_{nj} u_j,
\end{aligned}$$

where

$$a_{nj} = \begin{cases} k_{n-j}, & j \leq n; \\ 0, & j > n. \end{cases}$$

If $(M, \lambda_n) \subseteq (M, \mu_n)$, then the infinite matrix (a_{nj}) is regular. So, appealing to Theorem 1.1,

$$\sup_{n\geq 0}\sum_{j=0}^{\infty}|a_{nj}| < \infty,$$

$$i.e., \sup_{n\geq 0}\sum_{j=0}^{n}|a_{nj}| < \infty,$$

$$i.e., \sup_{n\geq 0}\sum_{j=0}^{n}|k_{n-j}| < \infty,$$

$$i.e., \sup_{n\geq 0}\sum_{j=0}^{n}|k_j| < \infty,$$

$$i.e., \sum_{j=0}^{\infty}|k_j| < \infty.$$

Also, $\lim_{n\to\infty}\sum_{j=0}^{\infty}a_{nj} = 1$ implies that $\sum_{n=0}^{\infty}k_n = 1$. Conversely, if $\sum_{n=0}^{\infty}|k_n| < \infty$ and $\sum_{n=0}^{\infty}k_n = 1$, then it follows that (a_{nj}) is regular and so $\lim_{j\to\infty}u_j = t$ implies that $\lim_{n\to\infty}v_n = t$. Thus, $(M, \lambda_n) \subseteq (M, \mu_n)$. The proof of the theorem is now complete. □

Consequently, we have the following result.

Theorem 3.8 (Equivalence theorem) *The methods (M, λ_n), (M, μ_n) are equivalent, i.e., $(M, \lambda_n) \subseteq (M, \mu_n)$ and vice versa if and only if*

$$\sum_{n=0}^{\infty}|k_n| < \infty, \sum_{n=0}^{\infty}|h_n| < \infty;$$

and

$$\sum_{n=0}^{\infty}k_n = 1, \sum_{n=0}^{\infty}h_n = 1,$$

where

$$\frac{\mu(x)}{\lambda(x)} = k(x) = \sum_{n=0}^{\infty}k_n x^n,$$

$$\frac{\lambda(x)}{\mu(x)} = h(x) = \sum_{n=0}^{\infty}h_n x^n.$$

3.3 The Abel Method and the (M, λ_n) Method

We recall the following (see [11]).

Definition 3.6 The sequence $\{s_n\}$ is Abel summable to s if

$$\lim_{x \to 1-} (1-x) \sum_{n=0}^{\infty} s_n x^n \text{ exists and } = s.$$

The following result is well known (see [11]).

Theorem 3.9 *The Abel method is regular.*

Remark 3.4 It is worthwhile to note that the Abel method cannot be described by an infinite matrix.

The next result gives the connection between the Abel method and the Natarajan method (see [7]).

Theorem 3.10 ([7, Theorem 4.2]) *If $\{a_n\}$ is (M, λ_n)-summable to s, where (M, λ_n) is regular, then $\{a_n\}$ is Abel summable to s.*

Proof Let $\{u_n\}$ be the (M, λ_n)-transform of the sequence $\{a_k\}$ so that

$$u_n = \lambda_0 a_n + \lambda_1 a_{n-1} + \cdots + \lambda_n a_0, \quad n = 0, 1, 2, \ldots.$$

Then $\lim_{n \to \infty} u_n = s$. Now,

$$\left(\sum_{n=0}^{\infty} \lambda_n x^n\right)\left(\sum_{n=0}^{\infty} a_n x^n\right) = \sum_{n=0}^{\infty} u_n x^n,$$
$$\left[(1-x)\left(\sum_{n=0}^{\infty} a_n x^n\right)\right]\left(\sum_{n=0}^{\infty} \lambda_n x^n\right) = (1-x)\left(\sum_{n=0}^{\infty} u_n x^n\right),$$
$$\left[(1-x)\left(\sum_{n=0}^{\infty} a_n x^n\right)\right](1-x)\left(\sum_{n=0}^{\infty} \lambda_n x^n\right)\left(\sum_{n=0}^{\infty} x^n\right) = (1-x)\left(\sum_{n=0}^{\infty} u_n x^n\right),$$
$$\left[(1-x)\left(\sum_{n=0}^{\infty} a_n x^n\right)\right](1-x)\left(\sum_{n=0}^{\infty} \Lambda_n x^n\right) = (1-x)\left(\sum_{n=0}^{\infty} u_n x^n\right), \tag{3.17}$$

where $\Lambda_n = \sum_{k=0}^{n} \lambda_k$, $n = 0, 1, 2, \ldots$. Taking limit as $x \to 1-$ in (3.17), we have,

$$\lim_{x \to 1-} (1-x)\left(\sum_{n=0}^{\infty} a_n x^n\right) = \lim_{x \to 1-} (1-x)\left(\sum_{n=0}^{\infty} u_n x^n\right), \tag{3.18}$$

noting that in view of Theorem 3.9, $\lim_{x \to 1-} (1-x)\left(\sum_{n=0}^{\infty} \Lambda_n x^n\right) = 1$, since $\lim_{n \to \infty} \Lambda_n = \sum_{n=0}^{\infty} \lambda_n = 1$, (M, λ_n) being regular. Since $\lim_{n \to \infty} u_n = s$, $\lim_{x \to 1-} (1-x)\left(\sum_{n=0}^{\infty} u_n x^n\right) = s$, using Theorem 3.9 once again. Thus, from (3.17), it follows that

$$\lim_{x \to 1-} (1-x)\left(\sum_{n=0}^{\infty} a_n x^n\right) = s,$$

i.e., $\{a_n\}$ is Abel summable to s, completing the proof of the theorem. □

Remark 3.5 The converse of Theorem 3.10 is not true. Consider the (M, λ_n) method with $\lambda_0 = \lambda_1 = \frac{1}{2}$, $\lambda_n = 0$, $n \geq 2$, i.e., the Y-method. The sequence $\{1, -1, 2, -2, 3, -3, \ldots\}$ is Abel summable to $\frac{1}{4}$ but not (M, λ_n) summable.

Remark 3.6 Theorem 3.10 implies that any Tauberian theorem for the Abel method is a Tauberian theorem for any regular (M, λ_n) method. It is worthwhile to have a Tauberian theorem for a regular (M, λ_n) method, which is not a Tauberian theorem for the Abel method.

We now recall that a product theorem means the following: given regular methods A, B, does $x = \{x_k\} \in (A)$ imply $B(x) \in (A)$, limits being the same, where (A) denotes the convergence field of A? i.e., does "$A(x)$ converges" imply "$A(B(x))$ converges to the same limit"?

In view of (3.18), we have the following product theorem.

Theorem 3.11 (Product theorem) *If $\{a_n\}$ is Abel summable to s, then $(M, \lambda_n)(\{a_n\})$ is also Abel summable to s.*

Proof If $\{a_n\}$ is Abel summable to s, using (3.18),

$$\lim_{x \to 1-} (1-x)\left(\sum_{n=0}^{\infty} u_n x^n\right) = s,$$

where $\{u_n\}$ is the (M, λ_n)-transform of $\{a_n\}$, completing the proof. □

3.4 The Euler Method and the (M, λ_n) Method

The Euler summability method is defined as follows (see [11, pp. 56–57]).

Definition 3.7 Let $r \in \mathbb{C} - \{1, 0\}$, $\mathbb{C}$ being the field of complex numbers. The Euler method of order r or the (E, r) method is defined by the infinite matrix $(e_{nk}^{(r)})$, where

$$e_{nk}^{(r)} = \begin{cases} {}^{n}C_{k} r^{k}(1-r)^{n-k}, & k \leq n; \\ 0, & k > n. \end{cases}$$

If $r = 1$,

$$e_{nk}^{(1)} = \begin{cases} 1, & k = n; \\ 0, & k \neq n. \end{cases}$$

If $r = 0$,

$$e_{nk}^{(0)} = 0, \; n = 0, 1, 2, \ldots; k = 1, 2, \ldots;$$
$$e_{n0}^{(0)} = 1, \; n = 0, 1, 2, \ldots.$$

The following result is well known (see [11, p. 57, Theorem 3.15]).

Theorem 3.12 *The* (E, r) *method is regular if and only if* r *is real and* $0 < r \leq 1$.

Remark 3.7 In the context of Theorem 3.12, we note that

$$\lim_{n\to\infty} e_{nk}^{(r)} = 0, \quad k = 0, 1, 2, \ldots$$

if and only if

$$|1 - r| < 1$$

(see [11, p. 57, proof of Theorem 3.15]).

Theorem 3.13 ([11, pp. 58–59, Corollary 3.17]) *If* $r \neq 0$, *the* (E, r) *matrix is invertible and its inverse is the* $(E, \frac{1}{r})$ *matrix.*

The following result is very useful (see [10, p. 234, Theorem 176]).

Theorem 3.14 *If* $\lim_{n\to\infty} a_n = 0$ *and* $\sum_{n=0}^{\infty} |b_n| < \infty$, *then*

$$\lim_{n\to\infty} (a_0 b_n + a_1 b_{n-1} + \cdots + a_n b_0) = 0.$$

We now prove an interesting product theorem involving the (E, r) and (M, λ_n) methods.

Theorem 3.15 ([8, Theorem 2.3]) *Given a sequence* $x = \{x_k\}$, *if* $(E, r)(x)$ *converges to* s, *then* $(E, r)((M, \lambda_n)(x))$ *converges to*

$$s\left(\lambda_0 + \sum_{n=1}^{\infty} \lambda_n r^{n-1}\right),$$

where we suppose that the (E, r) *method is regular.*

Proof Let

$$\tau_n = \sum_{k=0}^{n} {}^n c_k r^k (1-r)^{n-k} x_k, \tag{3.19}$$
$$t_n = \lambda_n x_0 + \lambda_{n-1} x_1 + \cdots + \lambda_0 x_n, \quad n = 0, 1, 2, \ldots.$$

By hypothesis, $\lim_{n\to\infty} \tau_n = s$ and $\sum_{n=0}^{\infty} |\lambda_n| < \infty$. Let $(E, r)(\{t_n\}) = \{\tau'_n\}$ so that

$$\begin{aligned}
\tau'_n &= \sum_{k=0}^{n} {}^n c_k r^k (1-r)^{n-k} t_k \\
&= (1-r)^n t_0 + {}^n c_1 r (1-r)^{n-1} t_1 + {}^n c_2 r^2 (1-r)^{n-2} t_2 + \cdots + r^n t_n \\
&= (1-r)^n (\lambda_0 x_0) + {}^n c_1 r (1-r)^{n-1} (\lambda_0 x_1 + \lambda_1 x_0) \\
&\quad + {}^n c_2 r^2 (1-r)^{n-2} (\lambda_0 x_2 + \lambda_1 x_1 + \lambda_2 x_0) + \cdots \\
&\quad + r^n (\lambda_0 x_n + \lambda_1 x_{n-1} + \cdots + \lambda_n x_0) \\
&= \lambda_0 [(1-r)^n x_0 + {}^n c_1 r (1-r)^{n-1} x_1 + {}^n c_2 r^2 (1-r)^{n-2} x_2 + \cdots + r^n x_n] \\
&\quad + \lambda_1 [{}^n c_1 r (1-r)^{n-1} x_0 + {}^n c_2 r^2 (1-r)^{n-2} x_1 + \cdots + r^n x_{n-1}] \\
&\quad + \lambda_2 [{}^n c_2 r^2 (1-r)^{n-2} x_0 + {}^n c_3 r^3 (1-r)^{n-3} x_1 + \cdots + r^n x_{n-2}] \\
&\quad + \cdots + \lambda_n r^n x_0 \\
&= \lambda_0 \left[\sum_{k=0}^{n} {}^n c_k r^k (1-r)^{n-k} x_k\right] + \lambda_1 \left[\sum_{k=1}^{n} {}^n c_k r^k (1-r)^{n-k} x_{k-1}\right] \\
&\quad + \lambda_2 \left[\sum_{k=2}^{n} {}^n c_k r^k (1-r)^{n-k} x_{k-2}\right] + \cdots + \lambda_n r^n x_0 \\
&= \lambda_0 \tau_n + \lambda_1 \left[\sum_{k=1}^{n} {}^n c_k r^k (1-r)^{n-k} x_{k-1}\right] \\
&\quad + \lambda_2 \left[\sum_{k=2}^{n} {}^n c_k r^k (1-r)^{n-k} x_{k-2}\right] + \cdots + \lambda_n r^n x_0.
\end{aligned} \tag{3.20}$$

Now,

$$\sum_{k=1}^{n} {}^{n}c_k r^k (1-r)^{n-k} x_{k-1}$$

$$= \sum_{j=0}^{n-1} {}^{n}c_{j+1} r^{j+1} (1-r)^{n-j-1} x_j$$

$$= \sum_{j=0}^{n-1} \left[{}^{n}c_{j+1} r^{j+1} (1-r)^{n-j-1} \left\{ \sum_{k=0}^{j} {}^{j}c_k \left(\frac{1}{r}\right)^k \left(1-\frac{1}{r}\right)^{j-k} \tau_k \right\} \right],$$

using (3.19) and Theorem 3.13

$$= \sum_{k=0}^{n-1} \left[r(1-r)^{n-k-1} \tau_k \left\{ \sum_{j=k}^{n-1} (-1)^{j-k}\, {}^{n}c_{j+1}\, {}^{j}c_k \right\} \right],$$

interchanging the order of summation. (3.21)

We now use the identity

$$\sum_{k=0}^{n-1} \left(\sum_{j=k}^{n-1} (-1)^{j-k}\, {}^{n}c_{j+1}\, {}^{j}c_k \right) z^k = \sum_{k=0}^{n-1} z^k,$$

to conclude that

$$\sum_{j=k}^{n-1} (-1)^{j-k}\, {}^{n}c_{j+1}\, {}^{j}c_k = 1, \quad 0 \le k \le n-1. \tag{3.22}$$

Using (3.21) and (3.22), we have,

$$\sum_{k=1}^{n} {}^{n}c_k r^k (1-r)^{n-k} x_{k-1} = \sum_{k=0}^{n-1} r(1-r)^{n-k-1} \tau_k. \tag{3.23}$$

Using (3.23) and similar results, (3.20) can now be written as

$$\tau'_n = \lambda_0 \tau_n + \lambda_1 \left[\sum_{k=0}^{n-1} r(1-r)^{n-k-1} \tau_k \right]$$

$$+ \lambda_2 \left[\sum_{k=0}^{n-2} r^2 (1-r)^{n-k-2} \tau_k \right] + \cdots + \lambda_n r^n \tau_0$$

$$= \lambda_0 (\tau_n - s) + \lambda_1 \left[\sum_{k=0}^{n-1} r(1-r)^{n-k-1} (\tau_k - s) \right]$$

$$
\begin{aligned}
&+ \lambda_2\left[\sum_{k=0}^{n-2} r^2(1-r)^{n-k-2}(\tau_k - s)\right] + \cdots + \lambda_n r^n(\tau_0 - s) \\
&+ s\left[\lambda_0 + \lambda_1\left\{\sum_{k=0}^{n-1} r(1-r)^{n-k-1}\right\}\right. \\
&\qquad \left. + \lambda_2\left\{\sum_{k=0}^{n-2} r^2(1-r)^{n-k-2}\right\} + \cdots + \lambda_n r^n\right] \\
= {}& \lambda_0(\tau_n - s) + \lambda_1\left[\sum_{k=0}^{n-1} r(1-r)^{n-k-1}(\tau_k - s)\right] \\
&+ \lambda_2\left[\sum_{k=0}^{n-2} r^2(1-r)^{n-k-2}(\tau_k - s)\right] + \cdots + \lambda_n r^n(\tau_0 - s) \\
&+ s\left[\lambda_0 + \lambda_1 r\left\{\frac{1-(1-r)^n}{1-(1-r)}\right\}\right. \\
&\qquad \left. + \lambda_2 r^2\left\{\frac{1-(1-r)^{n-1}}{1-(1-r)}\right\} + \cdots + \lambda_n r^n\right] \\
= {}& \lambda_0(\tau_n - s) + \lambda_1\left[\sum_{k=0}^{n-1} r(1-r)^{n-k-1}(\tau_k - s)\right] \\
&+ \lambda_2\left[\sum_{k=0}^{n-2} r^2(1-r)^{n-k-2}(\tau_k - s)\right] + \cdots + \lambda_n r^n(\tau_0 - s) \\
&+ s\left[\lambda_0 + \lambda_1\{1-(1-r)^n\}\right. \\
&\qquad \left. + \lambda_2 r\{1-(1-r)^{n-1}\} + \cdots + \lambda_n r^{n-1}\{1-(1-r)\}\right] \\
= {}& \lambda_0(\tau_n - s) + \lambda_1\left[\sum_{k=0}^{n-1} r(1-r)^{n-k-1}(\tau_k - s)\right] \\
&+ \lambda_2\left[\sum_{k=0}^{n-2} r^2(1-r)^{n-k-2}(\tau_k - s)\right] + \cdots + \lambda_n r^n(\tau_0 - s) \\
&+ s\left[(\lambda_0 + \lambda_1 + \lambda_2 r + \cdots + \lambda_n r^{n-1})\right. \\
&\qquad \left. - \{\lambda_1(1-r)^n + \lambda_2 r(1-r)^{n-1} + \cdots + \lambda_n r^{n-1}(1-r)\}\right]. \qquad (3.24)
\end{aligned}
$$

Since the (E, r) method is regular, r is real and $0 < r \leq 1$, using Theorem 3.12. So,

$$\sum_{n=0}^{\infty} |\lambda_n r^{n-1}| \leq \sum_{n=0}^{\infty} |\lambda_n| < \infty$$

and

$$|(1-r)^n| = |1-r|^n \to 0, \quad n \to \infty,$$

since $|1 - r| < 1$, in view of Remark 3.7. Using Theorem 3.14, we have,

$$\lim_{n\to\infty} \left[\lambda_1(1-r)^n + \lambda_2 r(1-r)^{n-1} + \cdots + \lambda_n r^{n-1}(1-r)\right] = 0.$$

Let $\alpha_n = \lambda_n r^n$. Then $\sum_{n=0}^{\infty} |\alpha_n| \le \sum_{n=0}^{\infty} |\lambda_n| < \infty$, since $0 < r \le 1$. Let

$$\beta_n = \sum_{k=0}^{n-1} (1-r)^{n-k-1}(\tau_k - s), \quad n = 0, 1, 2, \ldots, \beta_0 = 0.$$

$\lim_{n\to\infty} (\tau_n - s) = 0$ and $\sum_{n=0}^{\infty} |(1-r)^n| = \sum_{n=0}^{\infty} |1-r|^n < \infty$, since $|1 - r| < 1$, using Remark 3.7 again. Thus, $\lim_{n\to\infty} \beta_n = 0$, using Theorem 3.14 once more. Since $\sum_{n=0}^{\infty} |\alpha_n| < \infty$ and $\lim_{n\to\infty} \beta_n = 0$, appealing to Theorem 3.14 again, we see that

$$\lim_{n\to\infty} \left[\lambda_1 r \left\{\sum_{k=0}^{n-1} (1-r)^{n-k-1}(\tau_k - s)\right\} + \lambda_2 r^2 \left\{\sum_{k=0}^{n-2} (1-r)^{n-k-2}(\tau_k - s)\right\} + \cdots + \lambda_n r^n(\tau_0 - s)\right] = 0.$$

Taking limit as $n \to \infty$ in (3.24), we have,

$$\lim_{n\to\infty} \tau_n' = s\left[\lambda_0 + \sum_{k=1}^{\infty} \lambda_k r^{k-1}\right],$$

noting that the series on the right converges, since $\sum_{k=1}^{\infty} |\lambda_k r^{k-1}| \le \sum_{k=0}^{\infty} |\lambda_k| < \infty$. In other words, $(E, r)((M, \lambda_n)(x))$ converges to $s\left[\lambda_0 + \sum_{k=1}^{\infty} \lambda_k r^{k-1}\right]$, completing the proof of the theorem. □

Corollary 3.1 (Product theorem) *If we want to get the same limit s, we have to choose $\{\lambda_n\}$ such that*

$$\lambda_0 + \sum_{k=1}^{\infty} \lambda_k r^{k-1} = 1,$$

an example being the Y-method.

Corollary 3.2 *Any regular (E, r) method and (M, λ_n) method for which $\lambda_0 + \sum_{k=1}^{\infty} \lambda_k r^{k-1} = 1$ are consistent.*

In particular, any regular (E, r) method and the Y-method are consistent.

References

1. Peyerimhoff, A.: Lectures on Summability. Lecture Notes in Mathematics, vol. 107. Springer, Berlin (1969)
2. Natarajan, P.N.: A generalization of a theorem of Móricz and Rhoades on Weighted means. Comment. Math. Prace Mat. **52**, 29–37 (2012)
3. Hardy, G.H.: A theorem concerning summable series. Proc. Cambridge Philos. Soc. **20**, 304–307 (1920-21)
4. Móricz, F., Rhoades, B.E.: An equivalent reformulation of summability by weighted mean methods. revisited. Linear Algebra Appl. **349**, 187–192 (2002)
5. Natarajan, P.N.: Another theorem on weighted means. Comment. Math. Prace Mat. **50**, 175–181 (2010)
6. Maddox, I.J.: Elements of Functional Analysis. Cambridge University Press, Cambridge (1977)
7. Natarajan, P.N.: On the (M, λ_n) method of summability. Analysis (München) **33**, 51–56 (2013)
8. Natarajan, P.N.: A product theorem for the Euler and the Natarajan methods of summability. Analysis (München) **33**, 189–195 (2013)
9. Natarajan, P.N.: New properties of the Natarajan method of summability. Comment. Math. Prace Mat. **55**, 9–15 (2015)
10. Hardy, G.H.: Divergent Series. Oxford (1949)
11. Powell, R.E., Shah, S.M.: Summability Theory and Applications. Prentice-Hall of India (1988)

Chapter 4
More Properties of the (M, λ_n) Method and Cauchy Multiplication of Certain Summable Series

In this chapter, we study some more properties of the Natarajan method. Some results on the Cauchy multiplication of certain summable series are also proved. This chapter is divided into 4 sections. The first section is devoted to a study of the (M, λ_n) method. For instance, we prove that the set $\mathcal{M}$ of all (M, λ_n) methods is an ordered abelian semigroup and there exist infinite chains of (M, λ_n) methods. In the second section, we study iteration of (M, λ_n) methods. In the third section, we prove a few results on the Cauchy multiplication of (M, λ_n)-summable series, while in the last section, we prove a couple of results on the Cauchy multiplication of Euler summable series.

4.1 Some Nice Properties of the (M, λ_n) Method

In the present section, following Defranza [1], we record some nice properties of the Natarajan method. In this context, we recall that

$$\ell = \left\{ x = \{x_k\} : \sum_{k=0}^{\infty} |x_k| < \infty \right\}.$$

In the sequel, for convenience, we write (M, λ) for (M, λ_n). Let $\mathcal{M}$ denote the set of all (M, λ) methods.

We note the following.

Theorem 4.1 *For any method* $(M, \lambda) \in \mathcal{M}$, $(M, \lambda) \in (\ell, \ell)$.

Let $\ell((M, \lambda))$ denote the set of all sequences $x = \{x_k\}$ such that $(M, \lambda)(x) \in \ell$.

Definition 4.1 Given the methods $(M, \lambda), (M, \mu) \in \mathcal{M}$, we say that

$$(M, \mu) \text{ is } \ell - \text{stronger than } (M, \lambda)$$

P.N. Natarajan, *Classical Summability Theory*, DOI 10.1007/978-981-10-4205-8_4

if

$$\ell((M, \lambda)) \subseteq \ell((M, \mu));$$

$$(M, \mu) \text{ is strictly } \ell - \text{stronger than } (M, \lambda)$$

if

$$\ell((M, \lambda)) \subsetneqq \ell((M, \mu));$$

$$(M, \lambda) \text{ and } (M, \mu) \text{ are } \ell - \text{equivalent}$$

if

$$\ell((M, \lambda)) = \ell((M, \mu)).$$

Given the methods $(M, \lambda), (M, \mu) \in \mathcal{M}$, we formally define

$$\lambda(x) = \sum_{n=0}^{\infty} \lambda_n x^n, \quad \mu(x) = \sum_{n=0}^{\infty} \mu_n x^n$$

and

$$a(x) = \frac{\lambda(x)}{\mu(x)} = \sum_{n=0}^{\infty} a_n x^n;$$
$$b(x) = \frac{\mu(x)}{\lambda(x)} = \sum_{n=0}^{\infty} b_n x^n.$$

Following an argument similar to the one in [2, Theorem 18], we can prove the following results.

Theorem 4.2 *If $(M, \lambda) \in \mathcal{M}$, then the series $\sum_{n=0}^{\infty} \lambda_n x^n$ converges for $|x| < 1$.*

Theorem 4.3 *If $(M, \lambda), (M, \mu) \in \mathcal{M}$, then $\sum_{n=0}^{\infty} a_n x^n$, $\sum_{n=0}^{\infty} b_n x^n$ have positive radii of convergence and*

(i) $\lambda_n = a_n \mu_0 + a_{n-1}\mu_1 + \cdots + a_0 \mu_n$;
(ii) $\mu_n = b_n \lambda_0 + b_{n-1}\lambda_1 + \cdots + b_0 \lambda_n$, $n = 0, 1, 2, \ldots$.

Further, if $s = \{s_n\} \in \ell((M, \lambda))$, then the series $s(x) = \sum_{n=0}^{\infty} s_n x^n$ has positive radius of convergence.

Given the sequences $\lambda = \{\lambda_n\}$, $\mu = \{\mu_n\}$, let $\lambda * \mu = \{g_n\}$ denote their Cauchy product, i.e.,

$$g_n = \sum_{k=0}^{n} \lambda_k \mu_{n-k}, \quad n = 0, 1, 2, \ldots.$$

Definition 4.2 Given $(M, \lambda), (M, \mu) \in \mathcal{M}$, we say that $(M, \lambda * \mu)$ is the symmetric product of (M, λ) and (M, μ).

Since $\sum_{n=0}^{\infty} |\lambda_n| < \infty$ and $\sum_{n=0}^{\infty} |\mu_n| < \infty$, in view of Abel's theorem on Cauchy multiplication of absolutely convergent series, it follows that $\sum_{n=0}^{\infty} |g_n| < \infty$ so that $(M, \lambda * \mu) \in \mathcal{M}$, whenever $(M, \lambda), (M, \mu) \in \mathcal{M}$.

We need the following result in the sequel.

Lemma 4.1 *For given sequences $\lambda = \{\lambda_n\}$, $\mu = \{\mu_n\}$, let $\gamma = \lambda * \mu$. Let the methods $(M, \lambda), (M, \gamma) \in \mathcal{M}$. Then,*

$$\ell((M, \lambda)) \subseteq \ell((M, \gamma))$$

if and only if

$$\mu \in \ell, \quad i.e., (M, \mu) \in \mathcal{M}.$$

Proof For any sequence $x = \{x_k\}$,

$$\begin{aligned}
((M, \gamma)(x))_n &= \gamma_n x_0 + \gamma_{n-1} x_1 + \cdots + \gamma_0 x_n \\
&= (\lambda_n \mu_0 + \lambda_{n-1} \mu_1 + \cdots + \lambda_0 \mu_n) x_0 \\
&\quad + (\lambda_{n-1} \mu_0 + \lambda_{n-2} \mu_1 + \cdots + \lambda_0 \mu_{n-1}) x_1 \\
&\quad + \cdots + (\lambda_0 \mu_0) x_n \\
&= (\lambda_n x_0 + \lambda_{n-1} x_1 + \cdots + \lambda_0 x_n) \mu_0 \\
&\quad + (\lambda_{n-1} x_0 + \lambda_{n-2} x_1 + \cdots + \lambda_0 x_{n-1}) \mu_1 \\
&\quad + \cdots + (\lambda_0 x_0) \mu_n \\
&= ((M, \lambda)(x))_n \mu_0 + ((M, \lambda)(x))_{n-1} \mu_1 \\
&\quad + \cdots + ((M, \lambda)(x))_0 \mu_n \\
&= \sum_{k=0}^{\infty} t_{nk} ((M, \lambda)(x))_k,
\end{aligned}$$

where

$$t_{nk} = \begin{cases} \mu_{n-k}, & k \leq n; \\ 0, & k > n. \end{cases}$$

In view of Theorem 2.2, $\ell((M,\lambda)) \subseteq \ell((M,\gamma))$ if and only if $\sum_{n=0}^{\infty} |\mu_n| < \infty$, i.e., $\mu \in \ell$, completing the proof. □

We now deduce an inclusion result.

Theorem 4.4 *Given the methods* (M,λ), $(M,\mu) \in \mathcal{M}$,

$$\ell((M,\lambda)) \subseteq \ell((M,\mu))$$

if and only if

$$b = \{b_n\} \in \ell.$$

Proof In Lemma 4.1, we replace the sequence μ by the sequence b so that

$$\gamma = \lambda * b = \mu$$

and so

$$\ell((M,\lambda)) \subseteq \ell((M,\mu))$$

if and only if $b = \{b_n\} \in \ell$. □

Corollary 4.1 *Let* (M,λ), $(M,\mu) \in \mathcal{M}$. *Then,*

(i) $\ell((M,\lambda)) = \ell((M,\mu))$ *if and only if* $\{a_n\} \in \ell$ *and* $\{b_n\} \in \ell$;
(ii) $\ell((M,\lambda)) \subsetneqq \ell((M,\mu))$ *if and only if* $\{a_n\} \notin \ell$ *and* $\{b_n\} \in \ell$.

Corollary 4.2 *For* $(M,\lambda) \in \mathcal{M}$, *let* $h(x) = \frac{1}{\lambda(x)}$. *Then,*

$$\ell((M,\lambda)) = \ell$$

if and only if $\{h_n\} \in \ell$.

Proof Let I be the identity map so that $\ell(I) = \ell$. Now, $I(x) = \sum_{n=0}^{\infty} i_n x^n = 1$, i.e., $i_0 = 1$, $i_n = 0$, $n \geq 1$. So $a(x) = \frac{\lambda(x)}{I(x)} = \lambda(x)$ and $b(x) = \frac{I(x)}{\lambda(x)} = \frac{1}{\lambda(x)} = h(x)$. Using Theorem 4.4, it follows that $\{h_n\} \in \ell$. □

Corollary 4.3 *Let* (M,λ), $(M,\mu) \in \mathcal{M}$ *and* $\gamma = \lambda * \mu$. *Then,*

$$\ell((M,\lambda)) \subseteq \ell((M,\gamma)).$$

Theorem 4.5 *Let* $(M,\lambda) \in \mathcal{M}$. *Let* $\mu = \{\lambda_0', \lambda_1, \lambda_2, \ldots\}$, $\lambda_0' \neq \lambda_0$. *Then* $(M,\mu) \in \mathcal{M}$ *and*

$$\ell((M,\lambda)) \cap \ell((M,\mu)) = \ell.$$

Proof Since $\{\lambda_n\} \in \ell$, it follows that $\{\mu_n\} \in \ell$ so that the method $(M, \mu) \in \mathcal{M}$. Now, for any sequence $x = \{x_k\}$,

$$\begin{aligned}((M, \mu)(x))_n &= \mu_0 x_n + \mu_1 x_{n-1} + \cdots + \mu_n x_0 \\ &= \lambda_0' x_n + \lambda_1 x_{n-1} + \cdots + \lambda_n x_0 \\ &= (\lambda_0 x_n + \lambda_1 x_{n-1} + \cdots + \lambda_n x_0) + (\lambda_0' - \lambda_0)x_n \\ &= ((M, \lambda)(x))_n + (\lambda_0' - \lambda_0)x_n. \end{aligned} \tag{4.1}$$

Let $\{x_n\} \in \ell$. Since (M, λ), $(M, \mu) \in (\ell, \ell)$, using Theorem 4.1,

$$\{((M, \lambda)(x))_n\} \in \ell \text{ and } \{((M, \mu)(x))_n\} \in \ell$$

so that

$$\{x_n\} \in \ell((M, \lambda)) \cap \ell((M, \mu)).$$

Consequently,

$$\ell \subseteq \ell((M, \lambda)) \cap \ell((M, \mu)).$$

Let, now, $\{x_n\} \in \ell((M, \lambda)) \cap \ell((M, \mu))$. Then, $\{((M, \lambda)(x))_n\} \in \ell$ and $\{((M, \mu)(x))_n\} \in \ell$. Now, using (4.1), it follows that $\{x_n\} \in \ell$. Thus,

$$\ell((M, \lambda)) \cap \ell((M, \mu)) \subseteq \ell.$$

Consequently,

$$\ell((M, \lambda)) \cap \ell((M, \mu)) = \ell,$$

completing the proof of the theorem. □

We now recall that $\mathcal{M}$ denotes the set of all (M, λ) methods. We now prove that $\mathcal{M}$ is an ordered abelian semigroup, the order relation being the set inclusion between summability fields of the type $\ell((M, \lambda))$ and the binary operation being the symmetric product $*$.

Lemma 4.2 *If (M, λ), $(M, \mu) \in \mathcal{M}$ and $\gamma = \lambda * \mu$, then $(M, \gamma) \in \mathcal{M}$ and*

$$\ell((M, \lambda)) \cup \ell((M, \mu)) \subseteq \ell((M, \gamma)).$$

Proof Since $\sum_{n=0}^{\infty} |\lambda_n| < \infty$ and $\sum_{n=0}^{\infty} |\mu_n| < \infty$, $\sum_{n=0}^{\infty} |\gamma_n| < \infty$ so that $(M, \gamma) \in \mathcal{M}$ as noted earlier. Using Corollary 4.3, we have,

$$\ell((M, \lambda)) \cup \ell((M, \mu)) \subseteq \ell((M, \gamma)).$$

□

Lemma 4.3 *Let (M, λ), (M, μ), $(M, \gamma) \in \mathcal{M}$. Let $\theta = \lambda * \gamma$, $\varphi = \mu * \gamma$.*

(i) if $\ell((M, \lambda)) \subseteq \ell((M, \mu))$, then $\ell((M, \theta)) \subseteq \ell((M, \varphi))$;
(ii) if $\ell((M, \lambda)) \subsetneqq \ell((M, \mu))$, then $\ell((M, \theta)) \subsetneqq \ell((M, \varphi))$.

Proof Let $b(x) = \frac{\mu(x)}{\lambda(x)}$ and $c(x) = \frac{\varphi(x)}{\theta(x)}$. By Theorem 4.4, $\{b_n\} \in \ell$. We claim that $\{c_n\} \in \ell$. First, we note that

$$\theta(x) = \lambda(x)\gamma(x)$$

and

$$\varphi(x) = \mu(x)\gamma(x).$$

Now,

$$\begin{aligned}\sum_{n=0}^{\infty} c_n x^n = c(x) &= \frac{\varphi(x)}{\theta(x)}\\ &= \frac{\mu(x)\gamma(x)}{\lambda(x)\gamma(x)}\\ &= \frac{\mu(x)}{\lambda(x)}\\ &= b(x)\\ &= \sum_{n=0}^{\infty} b_n x^n.\end{aligned}$$

Thus,

$$c_n = b_n, \quad n = 0, 1, 2, \ldots$$

and so $\{c_n\} \in \ell$. Using Theorem 4.4 again, we have,

$$\ell((M, \theta)) \subseteq \ell((M, \varphi)).$$

The second part of the theorem follows using Corollary 4.1(ii). □

We now have

Theorem 4.6 *With respect to "strictly ℓ-weaker than" as the order relation and symmetric product * as the binary operation and symmetric product * as the binary operation, $\mathcal{M}$ is an ordered abelian semigroup.*

Lemma 4.4 *Let $(M, \lambda) \in \mathcal{M}$. Define $\lambda_{-1} = 0$, $\mu = \frac{\lambda_{n-1}+\lambda_n}{2}$, $n \geq 0$. Then $(M, \mu) \in \mathcal{M}$ and*

$$\ell((M, \lambda)) \subsetneqq \ell((M, \mu)).$$

Proof Since $\{\lambda_n\} \in \ell$, $\{\mu_n\} \in \ell$ so that $(M, \mu) \in \mathcal{M}$. Now,

$$\begin{aligned}\mu(x) &= \sum_{n=0}^{\infty} \mu_n x^n \\ &= \sum_{n=0}^{\infty} \frac{\lambda_{n-1} + \lambda_n}{2} x^n \\ &= \frac{1+x}{2} \lambda(x).\end{aligned}$$

It is now clear that

$$\ell((M, \lambda)) \subsetneqq \ell((M, \mu)),$$

using Corollary 4.1(ii). □

We conclude this section with the following interesting result.

Theorem 4.7 *There are infinite chains of (M, λ) methods from $\mathcal{M}$.*

Proof Let $\{\lambda_n^{(1)}\}$ be a sequence such that $\sum_{n=0}^{\infty} |\lambda_n^{(1)}| < \infty$. Let $\lambda^{(1)}(x) = \sum_{n=0}^{\infty} \lambda_n^{(1)} x^n$. Then, $(M, \lambda^{(1)}) \in \mathcal{M}$. For $n \geq 2$, define

$$\lambda^{(n)}(x) = \left(\frac{1+x}{2}\right)^{n-1} \sum_{k=0}^{\infty} \lambda_k^{(n-1)} x^k.$$

Then, $(M, \lambda^{(n)}) \in \mathcal{M}$, $n \geq 2$. Now, applying Lemma 4.4 repeatedly, we have,

$$\ell((M, \lambda^{(1)})) \subsetneqq \ell((M, \lambda^{(2)})) \subsetneqq \cdots \subsetneqq \ell((M, \lambda^{(n)})) \subsetneqq \cdots.$$

This completes the proof of the theorem. □

4.2 Iteration of (M, λ_n) Methods

In this section, we prove a few theorems on the iteration product of Natarajan methods (see [3]).

For convenience, we denote the methods (M, p_n), (M, q_n), and (M, t_n) by (M, p), (M, q), and (M, t), respectively.

Theorem 4.8 *Let (M, p), (M, t) be regular methods. Then $(M, p)(M, t)$ is also regular, where we define, for $x = \{x_k\}$,*

$$((M, t)(M, p))(x) = (M, t)((M, p)(x)).$$

Proof Let $\{\alpha_n\}$ be the (M, p)-transform of $x = \{x_k\}$ and $\{\beta_n\}$ be the $(M, t)(M, p)$-transform of $x = \{x_k\}$. Then,

$$\alpha_n = \sum_{k=0}^{n} p_k x_{n-k},$$
$$\beta_n = \sum_{k=0}^{n} t_k \alpha_{n-k}, \quad n = 0, 1, 2, \ldots.$$

Now,

$$\begin{aligned}
\beta_n &= \sum_{k=0}^{n} t_k \alpha_{n-k} \\
&= t_0\alpha_n + t_1\alpha_{n-1} + \cdots + t_n\alpha_0 \\
&= t_0\left[\sum_{k=0}^{n} p_k x_{n-k}\right] + t_1\left[\sum_{k=0}^{n-1} p_k x_{n-1-k}\right] \\
&\quad + \cdots + t_{n-1}[p_0x_1 + p_1x_0] + t_n[p_0x_0] \\
&= t_0[p_0x_n + p_1x_{n-1} + \cdots + p_nx_0] \\
&\quad + t_1[p_0x_{n-1} + p_1x_{n-2} + \cdots + p_{n-1}x_0] \\
&\quad + \cdots + t_{n-1}[p_0x_1 + p_1x_0] + t_n[p_0x_0] \\
&= (p_0t_0)x_n + (p_0t_1 + p_1t_0)x_{n-1} \\
&\quad + \cdots + (p_0t_n + p_1t_{n-1} + \cdots + p_nt_0)x_0 \\
&= \sum_{k=0}^{\infty} c_{nk}x_k,
\end{aligned}$$

where

$$c_{nk} = \begin{cases} \sum_{\gamma=0}^{k} p_\gamma x_{k-\gamma}, & k \le n; \\ 0, & k > n. \end{cases}$$

Now,

$$\begin{aligned}
\sum_{k=0}^{\infty} |c_{nk}| &= \sum_{k=0}^{n} |c_{nk}| \\
&= \sum_{k=0}^{n} \left| \sum_{\gamma=0}^{k} p_\gamma t_{k-\gamma} \right|
\end{aligned}$$

$$\begin{aligned}
&\le \sum_{k=0}^{n}\left(\sum_{\gamma=0}^{k}|p_\gamma||t_{k-\gamma}|\right)\\
&= |p_0||t_0| + \left(\sum_{\gamma=0}^{1}|p_\gamma||t_{1-\gamma}|\right)\\
&\quad + \left(\sum_{\gamma=0}^{2}|p_\gamma||t_{2-\gamma}|\right) + \cdots + \left(\sum_{\gamma=0}^{n}|p_\gamma||t_{n-\gamma}|\right)\\
&= |p_0||t_0| + (|p_0||t_1| + |p_1||t_0|)\\
&\quad + (|p_0||t_2| + |p_1||t_1| + |p_2||t_0|)\\
&\quad + \cdots + (|p_0||t_n| + |p_1||t_{n-1}| + \cdots + |p_n||t_0|)\\
&= |p_0|\left(\sum_{k=0}^{n}|t_k|\right) + |p_1|\left(\sum_{k=0}^{n-1}|t_k|\right)\\
&\quad + \cdots + |p_n|(|t_0|)\\
&\le \left(\sum_{k=0}^{\infty}|t_k|\right)\left(\sum_{k=0}^{\infty}|p_k|\right), n = 0, 1, 2, \ldots,
\end{aligned}$$

so that

$$\sup_{n\ge 0}\sum_{k=0}^{\infty}|c_{nk}| \le \left(\sum_{k=0}^{\infty}|t_k|\right)\left(\sum_{k=0}^{\infty}|p_k|\right) < \infty,$$

noting that $\sum_{k=0}^{\infty}|t_k| < \infty$ and $\sum_{k=0}^{\infty}|p_k| < \infty$.

Using $\lim_{n\to\infty} p_n = 0$, $\sum_{n=0}^{\infty}|t_n| < \infty$ and Theorem 3.14, it follows that

$$\lim_{n\to\infty} c_{nk} = 0, \quad k = 0, 1, 2, \ldots.$$

Now,

$$\begin{aligned}
\sum_{k=0}^{\infty} c_{nk} &= \sum_{k=0}^{n} c_{nk}\\
&= \sum_{k=0}^{n}\left(\sum_{\gamma=0}^{k} p_\gamma t_{k-\gamma}\right)\\
&= p_0\left(\sum_{k=0}^{n} t_k\right) + p_1\left(\sum_{k=0}^{n-1} t_k\right) + \cdots + p_n(t_0)
\end{aligned}$$

$$= p_0 T_n + p_1 T_{n-1} + \cdots + p_n T_0, \text{ where } T_n = \sum_{k=0}^{n} t_k, \ n = 0, 1, 2, \ldots$$

$$= p_0(T_n - 1) + p_0(T_{n-1} - 1) + \cdots + p_n(T_0 - 1) + P_n, \text{ where } P_n = \sum_{k=0}^{n} p_k, \ n = 0, 1, 2, \ldots.$$

Since (M, t), (M, p) are regular, $\lim_{n\to\infty} T_n = \lim_{n\to\infty} P_n = 1$, in view of Theorem 3.4. Using $\lim_{n\to\infty}(T_n - 1) = 0$, $\sum_{n=0}^{\infty} |p_n| < \infty$ and Theorem 3.14, we have,

$$\lim_{n\to\infty} [p_0(T_n - 1) + p_1(T_{n-1} - 1) + \cdots + p_n(T_0 - 1)] = 0.$$

Thus,

$$\lim_{n\to\infty} \sum_{k=0}^{\infty} c_{nk} = 0 + 1 = 1.$$

In view of Theorem 1.1, (c_{nk}) is regular. In other words, $(M, t)(M, p)$ is regular, completing the proof of the theorem. □

Theorem 4.9 *Let (M, p), (M, q), (M, t) be regular methods. Then*

$$(M, p) \subseteq (M, q)$$

if and only if

$$(M, t)(M, p) \subseteq (M, t)(M, q).$$

Proof We write

$$r'_n = \sum_{k=0}^{n} p_k t_{n-k}, \ r''_n = \sum_{k=0}^{n} q_k t_{n-k}, \quad n = 0, 1, 2, \ldots.$$

Let

$$r' = \{r'_n\}, r'' = \{r''_n\}, r'(x) = \sum_{n=0}^{\infty} r'_n x^n,$$

$$r''(x) = \sum_{n=0}^{\infty} r''_n x^n, \ p(x) = \sum_{n=0}^{\infty} p_n x^n, \ q(x) = \sum_{n=0}^{\infty} q_n x^n$$

and

$$t(x) = \sum_{n=0}^{\infty} t_n x^n.$$

Since (M, p), (M, q), and (M, t) are regular, $(M, t)(M, p)$, and $(M, t)(M, q)$ are regular too in view of Theorem 4.8. To prove the present theorem, it suffices to show that

$$\frac{r''(x)}{r'(x)} = \frac{q(x)}{p(x)}.$$

We first note that

$$r'(x) = p(x)t(x) \text{ and } r''(x) = q(x)t(x)$$

so that

$$\frac{r''(x)}{r'(x)} = \frac{q(x)}{p(x)}.$$

□

We now use Theorem 2.1 of [4] to arrive at the conclusion, thus completing the proof of the theorem.

In view of Theorem 2.1 of [4], we can reformulate Theorem 4.9 as follows:

Theorem 4.10 *For given regular methods (M, p), (M, q) and (M, t), the following statements are equivalent:*

(i) $(M, p) \subseteq (M, q)$;
(ii) $(M, t)(M, p) \subseteq (M, t)(M, q)$;

and

(iii) $\sum_{n=0}^{\infty} |k_n| < \infty$ *and* $\sum_{n=0}^{\infty} k_n = 1$,

where $\frac{q(x)}{p(x)} = k(x) = \sum_{n=0}^{\infty} k_n x^n$.

4.3 Cauchy Multiplication of (M, λ_n)-Summable Series

In this section, we prove a few results on the Cauchy multiplication of (M, λ_n)-summable series (see [5]).

Theorem 4.11 *If* $\sum_{k=0}^{\infty} |a_k| < \infty$ *and* $\{b_k\}$ *is* (M, λ_n)*-summable to* B*, then* $\{c_k\}$ *is* (M, λ_n)*-summable to* AB*, where* $c_n = \sum_{k=0}^{n} a_k b_{n-k}$, $n = 0, 1, 2, \ldots$ *and* $\sum_{k=0}^{\infty} a_k = A$.

Proof Let

$$
\begin{aligned}
t_n &= \lambda_0 b_n + \lambda_1 b_{n-1} + \cdots + \lambda_n b_0, \\
u_n &= \lambda_0 c_n + \lambda_1 c_{n-1} + \cdots + \lambda_n c_0, \quad n = 0, 1, 2, \ldots.
\end{aligned}
$$

By hypothesis, $\lim_{n\to\infty} t_n = B$. Now,

$$
\begin{aligned}
u_n &= \lambda_0 c_n + \lambda_1 c_{n-1} + \cdots + \lambda_n c_0 \\
&= \lambda_0 (a_0 b_n + a_1 b_{n-1} + \cdots + a_n b_0) \\
&\quad + \lambda_1 (a_0 b_{n-1} + a_1 b_{n-2} + \cdots + a_{n-1} b_0) \\
&\quad + \cdots + \lambda_n (a_0 b_0) \\
&= a_0 (\lambda_0 b_n + \lambda_1 b_{n-1} + \cdots + \lambda_n b_0) \\
&\quad + a_1 (\lambda_0 b_{n-1} + \lambda_1 b_{n-2} + \cdots + \lambda_{n-1} b_0) \\
&\quad + \cdots + a_n (\lambda_0 b_0) \\
&= a_0 t_n + a_1 t_{n-1} + \cdots + a_n t_0 \\
&= [a_0 (t_n - B) + a_1 (t_{n-1} - B) + \cdots a_n (t_0 - B)] \\
&\quad + B[a_0 + a_1 + \cdots + a_n].
\end{aligned}
$$

Since $\sum_{n=0}^{\infty} |a_n| < \infty$ and $\lim_{n\to\infty} (t_n - B) = 0$, using Theorem 3.14,

$$
\lim_{n\to\infty} [a_0 (t_n - B) + a_1 (t_{n-1} - B) + \cdots + a_n (t_0 - B)] = 0,
$$

so that

$$
\lim_{n\to\infty} u_n = B \left(\sum_{n=0}^{\infty} a_n \right) = AB,
$$

i.e., $\{c_k\}$ is (M, λ_n)-summable to AB, completing the proof. □

It is now easy to prove the following theorem on similar lines.

Theorem 4.12 *If* $\sum_{k=0}^{\infty} |a_k| < \infty$ *and* $\sum_{k=0}^{\infty} a_k = A$ *and* $\sum_{k=0}^{\infty} b_k$ *is* (M, λ_n)*-summable to* B*, then* $\sum_{k=0}^{\infty} c_k$ *is* (M, λ_n)*-summable to* AB*, where* $c_n = \sum_{k=0}^{n} a_k b_{n-k}$, $n = 0, 1, 2, \ldots$.

Following Mears [6], we prove the following result.

Theorem 4.13 *If* $\sum_{k=0}^{\infty} a_k$ *is* (M, λ_n)*-summable to* A, $\sum_{k=0}^{\infty} b_k$ *is* (M, μ_n)*-summable to* B, *where* $\sum_{k=0}^{\infty} |\beta_k - \beta_{k+1}| < \infty$, $\{\beta_n\}$ *is the* (M, μ_n)*-transform of* $\{B_k\}$, $B_k = \sum_{j=0}^{k} b_j$, $k = 0, 1, 2, \ldots$, *then* $\sum_{k=0}^{\infty} c_k$ *is* (M, γ_n)*-summable to* AB, *where* $c_n = \sum_{k=0}^{n} a_k b_{n-k}$, $\gamma_n = \sum_{k=0}^{n} \lambda_k \mu_{n-k}$, $n = 0, 1, 2, \ldots$.

Proof First, we note that since $\sum_{n=0}^{\infty} |\lambda_n| < \infty$ and $\sum_{n=0}^{\infty} |\mu_n| < \infty$, $\sum_{n=0}^{\infty} |\gamma_n| < \infty$ so that the method (M, γ_n) is defined. Let

$$A_n = \sum_{k=0}^{n} a_k, \quad B_n = \sum_{k=0}^{n} b_k, \quad C_n = \sum_{k=0}^{n} c_k, \quad n = 0, 1, 2, \ldots.$$

Let

$$\alpha_n = \sum_{k=0}^{n} \lambda_k A_{n-k}, \; \beta_n = \sum_{k=0}^{n} \mu_k B_{n-k}, \; \delta_n = \sum_{k=0}^{n} \gamma_k C_{n-k}, \quad n = 0, 1, 2, \ldots.$$

We now claim that

$$\delta_n = \sum_{k=0}^{n} \alpha_k \beta_{n-k} - \sum_{k=0}^{n-1} \alpha_k \beta_{n-k-1}.$$

We first note that

$$C_n = a_0 B_n + a_1 B_{n-1} + \cdots + a_n B_0,$$

so that

$$\begin{aligned}
\delta_n &= \gamma_0 C_n + \gamma_1 C_{n-1} + \cdots + \gamma_n C_0 \\
&= \gamma_0 (a_0 B_n + a_1 B_{n-1} + \cdots + a_n B_0) \\
&\quad + \gamma_1 (a_0 B_{n-1} + a_1 B_{n-2} + \cdots + a_{n-1} B_0) \\
&\quad + \cdots + \gamma_n (a_0 B_0) \\
&= a_0 (\gamma_0 B_n + \gamma_1 B_{n-1} + \cdots + \gamma_n B_0) \\
&\quad + a_1 (\gamma_0 B_{n-1} + \gamma_1 B_{n-2} + \cdots + \gamma_{n-1} B_0) \\
&\quad + \cdots + a_n (\gamma_0 B_0).
\end{aligned}$$

One can prove that

$$\gamma_0 B_n + \gamma_1 B_{n-1} + \cdots + \gamma_n B_0 = \lambda_0 \beta_n + \lambda_1 \beta_{n-1} + \cdots + \lambda_n \beta_0, \quad n = 0, 1, 2, \ldots,$$

so that

$$\begin{aligned}
\delta_n &= a_0(\lambda_0\beta_n + \lambda_1\beta_{n-1} + \cdots + \lambda_n\beta_0) \\
&\quad + a_1(\lambda_0\beta_{n-1} + \lambda_1\beta_{n-2} + \cdots + \lambda_{n-1}\beta_0) \\
&\quad + \cdots + a_n(\lambda_0\beta_0) \\
&= A_0(\lambda_0\beta_n + \lambda_1\beta_{n-1} + \cdots + \lambda_n\beta_0) \\
&\quad + (A_1 - A_0)(\lambda_0\beta_{n-1} + \lambda_1\beta_{n-2} + \cdots + \lambda_{n-1}\beta_0) \\
&\quad + (A_2 - A_1)(\lambda_0\beta_{n-2} + \lambda_1\beta_{n-3} + \cdots + \lambda_{n-2}\beta_0) \\
&\quad + \cdots + (A_n - A_{n-1})(\lambda_0\beta_0) \\
&= \beta_n(\lambda_0 A_0) + \beta_{n-1}[\lambda_1 A_0 + \lambda_0(A_1 - A_0)] \\
&\quad + \beta_{n-2}[\lambda_2 A_0 + \lambda_1(A_1 - A_0) + \lambda_0(A_2 - A_1)] \\
&\quad + \cdots + \beta_0[\lambda_n A_0 + \lambda_{n-1}(A_1 - A_0) + \lambda_{n-2}(A_2 - A_1) \\
&\qquad\qquad + \cdots + \lambda_0(A_n - A_{n-1})] \\
&= \beta_n[\lambda_0 A_0] + \beta_{n-1}[(\lambda_0 A_1 + \lambda_1 A_0) - \lambda_0 A_0] \\
&\quad + \beta_{n-2}[(\lambda_0 A_2 + \lambda_1 A_1 + \lambda_2 A_0) - (\lambda_0 A_1 + \lambda_1 A_0)] \\
&\quad + \cdots + \beta_0[(\lambda_0 A_n + \lambda_1 A_{n-1} + \cdots + \lambda_n A_0) \\
&\qquad\qquad - (\lambda_0 A_{n-1} + \lambda_1 A_{n-2} + \cdots + \lambda_{n-1} A_0)] \\
&= \beta_n\alpha_0 + \beta_{n-1}[\alpha_1 - \alpha_0] + \beta_{n-2}[\alpha_2 - \alpha_1] \\
&\qquad\qquad + \cdots + \beta_0[\alpha_n - \alpha_{n-1}] \\
&= (\alpha_0\beta_n + \alpha_1\beta_{n-1} + \cdots + \alpha_n\beta_0) \\
&\qquad\qquad - (\alpha_0\beta_{n-1} + \alpha_1\beta_{n-2} + \cdots + \alpha_{n-1}\beta_0) \\
&= \sum_{k=0}^{n} \alpha_k\beta_{n-k} - \sum_{k=0}^{n-1} \alpha_k\beta_{n-k-1},
\end{aligned}$$

proving our claim. Thus, for $n = 0, 1, 2, \ldots,$

$$\begin{aligned}
\delta_n &= \sum_{k=0}^{n}(\alpha_k - A)\beta_{n-k} - \sum_{k=0}^{n-1}(\alpha_k - A)\beta_{n-k-1} \\
&\quad + A\left[\sum_{k=0}^{n}\beta_{n-k} - \sum_{k=0}^{n-1}\beta_{n-k-1}\right] \\
&= \sum_{k=0}^{n}(\alpha_k - A)\beta_{n-k} - \sum_{k=0}^{n-1}(\alpha_k - A)\beta_{n-k-1} + A\beta_n
\end{aligned}$$

$$= \sum_{k=0}^{n-1}(\alpha_k - A)(\beta_{n-k} - \beta_{n-k-1}) + (\alpha_n - A)\beta_0 + A\beta_n.$$

Since $\lim_{k\to\infty}(\alpha_k - A) = 0$ and $\sum_{k=0}^{\infty}|\beta_k - \beta_{k+1}| < \infty$, in view of Theorem 3.14,

$$\lim_{n\to\infty}\sum_{k=0}^{n-1}(\alpha_k - A)(\beta_{n-k} - \beta_{n-k-1}) = 0,$$

so that

$$\lim_{n\to\infty}\delta_n = AB,$$

i.e., $\sum_{k=0}^{\infty}c_k$ is (M, γ_n)-summable to AB, completing the proof of the theorem. □

Remark 4.1 If (M, λ_n), (M, μ_n) are regular, then $\sum_{n=0}^{\infty}\lambda_n = \sum_{n=0}^{\infty}\mu_n = 1$, in view of Theorem 3.4. By Abel's theorem on Cauchy multiplication of absolutely convergent series,

$$\sum_{n=0}^{\infty}\gamma_n = \left(\sum_{n=0}^{\infty}\lambda_n\right)\left(\sum_{n=0}^{\infty}\mu_n\right) = 1.$$

Again appealing to Theorem 3.4, it follows that (M, γ_n) is regular too.

4.4 Cauchy Multiplication of Euler Summable Series

Theorem 4.14 $\sum_{k=0}^{\infty}|x_k| < \infty$ *and* $\{y_k\}$ *is* (E, r)*-summable to* σ*, where the* (E, r) *method is regular, then* $\{z_k\}$ *is* (E, r)*-summable to*

$$\sigma\left[x_0 + \sum_{k=1}^{\infty}x_k r^{k-1}\right],$$

where $z_n = \sum_{k=0}^{n}x_k y_{n-k}$, $n = 0, 1, 2, \ldots$.

Proof Let $\{\sigma_n\}$ be the (E, r)-transform of $\{y_k\}$. Then,

$$\sigma_n = \sum_{k=0}^{n} {}^n c_k r^k (1-r)^{n-k} y_k, \quad n = 0, 1, 2, \ldots. \tag{4.2}$$

By hypothesis, $\lim_{n\to\infty} \sigma_n = \sigma$. Let $\{\tau_n\}$ be the (E, r)-transform of $\{z_k\}$ so that

$$\begin{aligned}
\tau_n &= \sum_{k=0}^{n} {}^n c_k r^k (1-r)^{n-k} z_k \\
&= (1-r)^n z_0 + {}^n c_1 r (1-r)^{n-1} z_1 + {}^n c_2 r^2 (1-r)^{n-2} z_2 \\
&\qquad\qquad + \cdots + r^n z_n \\
&= (1-r)^n (x_0 y_0) + {}^n c_1 r (1-r)^{n-1} (x_0 y_1 + x_1 y_0) \\
&\quad + {}^n c_2 r^2 (1-r)^{n-2} (x_0 y_2 + x_1 y_1 + x_2 y_0) \\
&\quad + \cdots + r^n (x_0 y_n + x_1 y_{n-1} + \cdots + x_n y_0) \\
&= x_0 [(1-r)^n y_0 + {}^n c_1 r (1-r)^{n-1} y_1 + \cdots + r^n y_n] \\
&\quad + x_1 [{}^n c_1 r (1-r)^{n-1} y_0 + {}^n c_2 r^2 (1-r)^{n-2} y_1 + \cdots + r^n y_{n-1}] \\
&\quad + \cdots + x_n [r^n y_0] \\
&= x_0 \left[\sum_{k=0}^{n} {}^n c_k r^k (1-r)^{n-k} y_k \right] \\
&\quad + x_1 \left[\sum_{k=1}^{n} {}^n c_k r^k (1-r)^{n-k} y_{k-1} \right] \\
&\quad + x_2 \left[\sum_{k=2}^{n} {}^n c_k r^k (1-r)^{n-k} y_{k-2} \right] + \cdots + x_n [r^n y_0] \\
&= x_0 \sigma_n + x_1 \left[\sum_{k=1}^{n} {}^n c_k r^k (1-r)^{n-k} y_{k-1} \right] \\
&\quad + x_2 \left[\sum_{k=2}^{n} {}^n c_k r^k (1-r)^{n-k} y_{k-2} \right] + \cdots + x_n [r^n y_0].
\end{aligned} \tag{4.3}$$

Now,

$$\begin{aligned}
\sum_{k=1}^{n} {}^n c_k r^k (1-r)^{n-k} y_{k-1} &= \sum_{j=0}^{n-1} {}^n c_{j+1} r^{j+1} (1-r)^{n-j-1} y_j \\
&= \sum_{j=0}^{n-1} \left[{}^n c_{j+1} r^{j+1} (1-r)^{n-j-1} \right.
\end{aligned}$$

$$\times \left\{ \sum_{k=0}^{j} {}^{j}c_k \left(\frac{1}{r}\right)^k \left(1 - \frac{1}{r}\right)^{j-k} \sigma_k \right\}\Bigg],$$

using Theorem 3.13 and (4.2)

$$= \sum_{k=0}^{n-1} \left[r(1-r)^{n-k-1} \sigma_k \left\{ \sum_{j=k}^{n-1} (-1)^{j-k}\, {}^{n}c_{j+1}\, {}^{j}c_k \right\} \right],$$

interchanging the order summation. (4.4)

Using the identity

$$\sum_{k=0}^{n-1} \left(\sum_{j=k}^{n-1} (-1)^{j-k}\, {}^{n}c_{j+1}\, {}^{j}c_k \right) z^k = \sum_{k=0}^{n-1} z^k,$$

we have,

$$\sum_{j=k}^{n-1} (-1)^{j-k}\, {}^{n}c_{j+1}\, {}^{j}c_k = 1, \quad 0 \le k \le n-1. \tag{4.5}$$

Thus, using (4.4) and (4.5), we have,

$$\sum_{k=1}^{n} {}^{n}c_k r^k (1-r)^{n-k} y_{k-1} = \sum_{k=0}^{n-1} r(1-r)^{n-k-1} \sigma_k. \tag{4.6}$$

Using (4.6) and similar results, (4.3) can now be written as

$$\begin{aligned}
\tau_n &= x_0 \sigma_n + x_1 \left[\sum_{k=0}^{n-1} r(1-r)^{n-k-1} \sigma_k \right] \\
&\quad + x_2 \left[\sum_{k=0}^{n-2} r^2 (1-r)^{n-k-2} \sigma_k \right] + \cdots + x_n [r^n \sigma_0] \\
&= x_0(\sigma_n - \sigma) + x_1 \left[\sum_{k=0}^{n-1} r(1-r)^{n-k-1} (\sigma_k - \sigma) \right] \\
&\quad + x_2 \left[\sum_{k=0}^{n-2} r^2 (1-r)^{n-k-2} (\sigma_k - \sigma) \right] + \cdots + x_n [r^n (\sigma_0 - \sigma)] \\
&\quad + \sigma \left[x_0 + x_1 \left\{ \sum_{k=0}^{n-1} r(1-r)^{n-k-1} \right\} \right.
\end{aligned}$$

$$
\begin{aligned}
&+x_2\left\{\sum_{k=0}^{n-2} r^2(1-r)^{n-k-2}\right\}+\cdots+x_n\{r^n\}\Bigg]\\
&= x_0(\sigma_n-\sigma)+x_1\left[\sum_{k=0}^{n-1} r(1-r)^{n-k-1}(\sigma_k-\sigma)\right]\\
&\quad + x_2\left[\sum_{k=0}^{n-2} r^2(1-r)^{n-k-2}(\sigma_k-\sigma)\right]+\cdots+x_n[r^n(\sigma_0-\sigma)]\\
&\quad + \sigma\left[x_0+x_1 r\left\{\frac{1-(1-r)^n}{1-(1-r)}\right\}+x_2 r^2\left\{\frac{1-(1-r)^{n-1}}{1-(1-r)}\right\}+\cdots+x_n r^n\right]\\
&= x_0(\sigma_n-\sigma)+x_1\left[\sum_{k=0}^{n-1} r(1-r)^{n-k-1}(\sigma_k-\sigma)\right]\\
&\quad + x_2\left[\sum_{k=0}^{n-2} r^2(1-r)^{n-k-2}(\sigma_k-\sigma)\right]+\cdots+x_n[r^n(\sigma_0-\sigma)]\\
&\quad + \sigma[x_0+x_1\{1-(1-r)^n\}+x_2 r\{1-(1-r)^{n-1}\}\\
&\qquad\qquad +\cdots+x_n r^{n-1}\{1-(1-r)\}]\\
&= x_0(\sigma_n-\sigma)+x_1 r\left\{\sum_{k=0}^{n-1}(1-r)^{n-k-1}(\sigma_k-\sigma)\right\}\\
&\quad + x_2 r^2\left\{\sum_{k=0}^{n-2}(1-r)^{n-k-2}(\sigma_k-\sigma)\right\}+\cdots+x_n r^n\{\sigma_0-\sigma\}\\
&\quad + \sigma[(x_0+x_1+x_2 r+\cdots+x_n r^{n-1})\\
&\qquad\qquad - \{x_1(1-r)^n+x_2 r(1-r)^{n-1}+\cdots+x_n r^{n-1}(1-r)\}].
\end{aligned}
\tag{4.7}
$$

Since the (E, r) method is regular, $0 < r \leq 1$, using Theorem 3.12. Now,

$$
\begin{aligned}
\sum_{n=1}^{\infty} |x_n r^{n-1}| &\leq \sum_{n=0}^{\infty} |x_n|, \quad \text{since } 0 < r \leq 1\\
&< \infty
\end{aligned}
$$

and $|(1-r)^n| = |1-r|^n \to 0, n \to \infty$, using Remark 3.7, since the (E, r) method is regular. Note that the sequence

$$
\{x_1(1-r)^n + x_2 r(1-r)^{n-1} + \cdots + x_n r^{n-1}(1-r)\}
$$

is the Cauchy product of the sequences $\{x_n r^{n-1}\}$ and $\{(1-r)^n\}$. In view of Theorem 3.14,

$$\lim_{n\to\infty} \{x_1(1-r)^n + x_2 r(1-r)^{n-1} + \cdots + x_n r^{n-1}(1-r)\} = 0.$$

Let $\alpha_n = x_n r^n$. $\sum_{n=0}^{\infty} |\alpha_n| = \sum_{n=0}^{\infty} |x_n r^n| \le \sum_{n=0}^{\infty} |x_n| < \infty$, since $0 < r \le 1$, using Theorem 3.12 again. Let

$$\beta_n = \sum_{k=0}^{n-1} (1-r)^{n-k-1}(\sigma_k - \sigma).$$

Note that $\{\beta_n\}$ is the Cauchy product of the sequences $\{(1-r)^n\}$ and $\{\sigma_n - \sigma\}$, where $\lim_{n\to\infty} (\sigma_n - \sigma) = 0$ and

$$\sum_{n=0}^{\infty} |(1-r)^n| = \sum_{n=0}^{\infty} |(1-r)|^n < \infty, \quad \text{since } |1-r| < 1,$$

in view of Remark 3.7. Using Theorem 3.14,

$$\lim_{n\to\infty} \beta_n = 0.$$

Since $\sum_{n=0}^{\infty} |\alpha_n| < \infty$ and $\lim_{n\to\infty} \beta_n = 0$, appealing to Theorem 3.14 once again, it follows that

$$\lim_{n\to\infty} \left[x_1 r \left\{ \sum_{k=0}^{n-1} (1-r)^{n-k-1}(\sigma_k - \sigma) \right\} + x_2 r^2 \left\{ \sum_{k=0}^{n-2} (1-r)^{n-k-2}(\sigma_k - \sigma) \right\} + \cdots + x_n r^n (\sigma_0 - \sigma) \right] = 0.$$

Thus, taking limit as $n \to \infty$ in (4.7), we have,

$$\lim_{n\to\infty} \tau_n = \sigma \left[x_0 + \sum_{k=1}^{\infty} x_k r^{k-1} \right],$$

where we note that the series on the right-hand side converges, since,

$$\sum_{k=1}^{\infty} |x_k r^{k-1}| \le \sum_{k=0}^{\infty} |x_k| < \infty,$$

$0 < r \leq 1$, the (E, r) method being regular, using Theorem 3.12. In other words, $\{z_k\}$ is (E, r)-summable to

$$\sigma\left[x_0 + \sum_{k=1}^{\infty} x_k r^{k-1}\right],$$

completing the proof of the theorem. □

The following result can be proved in a similar fashion.

Theorem 4.15 *If* $\sum_{k=0}^{\infty} |x_k| < \infty$ *and* $\sum_{k=0}^{\infty} y_k$ *is* (E, r)*-summable to* σ*, where the* (E, r) *method is regular, then* $\sum_{k=0}^{\infty} z_k$ *is* (E, r)*-summable to* $\sigma\left[x_0 + \sum_{k=1}^{\infty} x_k r^{k-1}\right]$*,* $z_n = \sum_{k=0}^{n} x_k y_{n-k}$*,* $n = 0, 1, 2, \ldots$.

References

1. Defranza, J.: An ordered set of Nörlund means. Int. J. Math. Math. Sci. **4**, 353–364 (1981)
2. Natarajan, P.N.: New properties of the Natarajan method of summability. Comment. Math. Prace Mat. **55**, 9–15 (2015)
3. Natarajan, P.N.: Iteration product of Natarajan methods of summability. Int. J. Phys. Math. Sci. **3**, 25–29 (2013)
4. Natarajan, P.N.: On the (M, λ_n) method of summability. Analysis (München) **33**, 51–56 (2013)
5. Natarajan, P.N.: Cauchy multiplication of (M, λ_n) summable series. Adv. Dev. Math. Sci. **3**, 39–46 (2012)
6. Mears, F.M.: Some multiplication theorems for the Nörlund means. Bull. Amer. Math. Soc. **41**, 875–880 (1935)

Chapter 5
The Silverman–Toeplitz, Schur, and Steinhaus Theorems for Four-Dimensional Infinite Matrices

In this chapter, we prove the Silverman–Toeplitz, Schur and Steinhaus theorems for four-dimensional infinite matrices. This chapter is divided into 3 sections. In the first section, we introduce a new definition of convergence of a double sequence and a double series and study some of its properties. In the second section, we prove the Silverman–Toeplitz theorem for four-dimensional infinite matrices, while, in the last section, we establish the Schur and Steinhaus theorems for such matrices.

5.1 A New Definition of Convergence of a Double Sequence and a Double Series

In this section, we introduce a new definition of limit of a double sequence and a double series and record a few results on convergent double sequences. The motivation for the introduction of this definition in classical analysis is the paper of Natarajan and Srinivasan [1], in which the authors had already introduced this definition in the context of a non-archimedean field. In this context, it is worthwhile to refer to the papers of Kojima [2] and Robison [3]. For the concepts and theorems presented in this chapter, one can refer to the papers of Natarajan [4, 5].

Definition 5.1 Let $\{x_{m,n}\}$ be a double sequence. We say that

$$\lim_{m+n\to\infty} x_{m,n} = x,$$

if for every $\epsilon > 0$, the set $\{(m, n) \in \mathbb{N}^2 : |x_{m,n} - x| \geq \epsilon\}$ is finite, $\mathbb{N}$ being the set of positive integers. In such a case, x is unique and x is called the "limit" of $\{x_{m,n}\}$. We also say that $\{x_{m,n}\}$ converges to x.

P.N. Natarajan, *Classical Summability Theory*, DOI 10.1007/978-981-10-4205-8_5

Definition 5.2 Let $\{x_{m,n}\}$ be a double sequence. We say that

$$\sum_{m,n=0}^{\infty,\infty} x_{m,n} = s,$$

if

$$\lim_{m+n\to\infty} s_{m,n} = s,$$

where

$$s_{m,n} = \sum_{k,\ell=0}^{m,n} x_{k,\ell}, \quad m, n = 0, 1, 2, \ldots.$$

In such a case, we say that the double series $\sum_{m,n=0}^{\infty,\infty} x_{m,n}$ converges to s.

Definition 5.3 The double series $\sum_{m,n=0}^{\infty,\infty} x_{m,n}$ is said to converge absolutely if $\sum_{m,n=0}^{\infty,\infty} |x_{m,n}|$ converges.

Remark 5.1 If $\sum_{m,n=0}^{\infty,\infty} x_{m,n}$ converges absolutely, it converges. However, the converse is false.

Remark 5.2 If $\lim_{m+n\to\infty} x_{m,n} = x$, then the double sequence $\{x_{m,n}\}$ is automatically bounded.

It is easy to prove the following results.

Theorem 5.1

$$\lim_{m+n\to\infty} x_{m,n} = x$$

if and only if

(i) $\lim_{m\to\infty} x_{m,n} = x$, $n = 0, 1, 2, \ldots$;
(ii) $\lim_{n\to\infty} x_{m,n} = x$, $m = 0, 1, 2, \ldots$;

and

(iii) for any $\epsilon > 0$, *there exists* $N \in \mathbb{N}$ *such that*

$$|x_{m,n} - x| < \epsilon, \quad m, n \geq N,$$

which we write as

$$\lim_{m,n\to\infty} x_{m,n} = x$$

(Note that this is Pringsheim's definition of limit of a double sequence).

Theorem 5.2 *If the double series* $\sum_{m,n=0}^{\infty,\infty} x_{m,n}$ *converges, then*

$$\lim_{m+n\to\infty} x_{m,n} = 0.$$

However, the converse is not true.

5.2 The Silverman–Toeplitz Theorem for Four-Dimensional Infinite Matrices

In this section, we prove Silverman–Toeplitz theorem for double sequences. We need the following definition in the sequel.

Definition 5.4 Given the four-dimensional infinite matrix $A = (a_{m,n,k,\ell})$, $m, n, k, \ell = 0, 1, 2, \ldots$ and a double sequence $\{x_{k,\ell}\}$, $k, \ell = 0, 1, 2, \ldots$, by the A-transform of $x = \{x_{k,\ell}\}$, we mean the double sequence $A(x) = \{(Ax)_{m,n}\}$, where

$$(Ax)_{m,n} = \sum_{k,\ell=0}^{\infty,\infty} a_{m,n,k,\ell} x_{k,\ell}, \quad m, n = 0, 1, 2, \ldots,$$

assuming that the series on the right converge. If $\lim_{m+n\to\infty} (Ax)_{m,n} = s$, we say that the double sequence $x = \{x_{k,\ell}\}$ is A-summable or summable A to s, written as

$$x_{k,\ell} \to s(A).$$

If $\lim_{m+n\to\infty} (Ax)_{m,n} = s$, whenever $\lim_{k+\ell\to\infty} x_{k,\ell} = s$, we say that the four-dimensional infinite matrix $A = (a_{m,n,k,\ell})$ is "regular."

We now prove the following important theorem.

Theorem 5.3 (Silverman–Toeplitz) *The four-dimensional infinite matrix* $A = (a_{m,n,k,\ell})$ *is regular if and only if*

$$\sup_{m,n\geq 0} \sum_{k,\ell=0}^{\infty,\infty} |a_{m,n,k,\ell}| < \infty; \tag{5.1}$$

$$\lim_{m+n\to\infty} a_{m,n,k,\ell} = 0, \quad k, \ell = 0, 1, 2, \ldots; \tag{5.2}$$

$$\lim_{m+n\to\infty} \sum_{k,\ell=0}^{\infty,\infty} a_{m,n,k,\ell} = 1; \tag{5.3}$$

$$\lim_{m+n\to\infty} \sum_{k=0}^{\infty} |a_{m,n,k,\ell}| = 0, \quad \ell = 0, 1, 2, \ldots; \tag{5.4}$$

and

$$\lim_{m+n\to\infty} \sum_{\ell=0}^{\infty} |a_{m,n,k,\ell}| = 0, \quad k = 0, 1, 2, \ldots. \tag{5.5}$$

Proof Proof of necessity part. Define the double sequence $\{x_{k,\ell}\}$ as follows: For any fixed $p, q = 0, 1, 2, \ldots$, let

$$x_{k,\ell} = \begin{cases} 1, & \text{if } k = p, \ell = q; \\ 0, & \text{otherwise.} \end{cases}$$

Then,

$$(Ax)_{m,n} = a_{m,n,p,q}.$$

Since $\lim\limits_{k+\ell\to\infty} x_{k,\ell} = 0$ and A is regular, it follows that (5.2) is necessary. Define the double sequence $\{x_{k,\ell}\}$, where $x_{k,\ell} = 1$, $k, \ell = 0, 1, 2, \ldots$. Now,

$$(Ax)_{m,n} = \sum_{k,\ell=0}^{\infty,\infty} a_{m,n,k,\ell}, \quad m, n = 0, 1, 2, \ldots,$$

where the double series on the right converges. Since $\lim\limits_{k+\ell\to\infty} x_{k,\ell} = 1$ and A is regular,

$$\lim_{m+n\to\infty} \sum_{k,\ell=0}^{\infty,\infty} a_{m,n,k,\ell} = 1,$$

so that (5.3) is necessary. We now prove that

$$\lim_{m+n\to\infty} \sum_{k=0}^{\infty} |a_{m,n,k,\ell}| = 0, \quad \ell = 0, 1, 2, \ldots.$$

Suppose not. Then, there exists $\ell_0 \in \mathbb{N}$ such that

$$\lim_{m+n\to\infty} \sum_{k=0}^{\infty} |a_{m,n,k,\ell_0}| = 0$$

does not hold. So, there exists an $\epsilon > 0$, such that

$$\left\{(m,n) \in \mathbb{N}^2 : \sum_{k=0}^{\infty} |a_{m,n,k,\ell_0}| > \epsilon\right\} \tag{5.6}$$

is infinite. Let us choose $m_0 = n_0 = r_0 = 1$. Choose $m_1, n_1 \in \mathbb{N}$ such that $m_1+n_1 > m_0 + n_0$,

$$\sum_{k=0}^{r_0} |a_{m_1,n_1,k,\ell_0}| < \frac{\epsilon}{8}, \text{ using (5.2)}$$

and

$$\sum_{k=0}^{\infty} |a_{m_1,n_1,k,\ell_0}| < \epsilon, \text{ using (5.6).}$$

We then choose $r_1 \in \mathbb{N}$ such that $r_1 > r_0$ and

$$\sum_{k=r_1+1}^{\infty} |a_{m_1,n_1,k,\ell_0}| < \frac{\epsilon}{8}, \text{ using (5.3).}$$

Inductively, we choose $m_p + n_p > m_{p-1} + n_{p-1}$ such that

$$\sum_{k=0}^{r_{p-1}} |a_{m_p,n_p,k,\ell_0}| < \frac{\epsilon}{8}; \tag{5.7}$$

$$\sum_{k=0}^{\infty} |a_{m_p,n_p,k,\ell_0}| > \epsilon \tag{5.8}$$

and then choose $r_p > r_{p-1}$ such that

$$\sum_{k=r_p+1}^{\infty} |a_{m_p,n_p,k,\ell_0}| < \frac{\epsilon}{8}. \tag{5.9}$$

In view of (5.7)–(5.9), we have,

$$\sum_{k=r_{p-1}+1}^{r_p} |a_{m_p,n_p,k,\ell_0}| > \epsilon - \frac{\epsilon}{8} - \frac{\epsilon}{8}$$
$$= \frac{3\epsilon}{4}.$$

We can now find $k_p \in \mathbb{N}$, $r_{p-1} < k_p \leq r_p$ such that

$$|a_{m_p,n_p,k_p,\ell_0}| > \frac{3\epsilon}{4}.$$

Define the double sequence $\{x_{k,\ell}\}$ as follows:

$$x_{k,\ell} = \begin{cases} sgn(a_{m_p,n_p,k_p,\ell_0}), & \text{if } \ell = \ell_0, k = k_p, p = 0, 1, 2, \ldots; \\ 0, & \text{if } \ell \neq \ell_0. \end{cases}$$

We first note that $\lim\limits_{k+\ell \to \infty} x_{k,\ell} = 0$. Now,

$$\left| \sum_{k=0}^{r_{p-1}} a_{m_p,n_p,k,\ell_0} x_{k,\ell_0} \right| \leq \sum_{k=0}^{r_{p-1}} |a_{m_p,n_p,k,\ell_0}|$$
$$< \frac{\epsilon}{8}, \quad \text{using (5.7);}$$

$$\left| \sum_{k=r_p+1}^{\infty} a_{m_p,n_p,k,\ell_0} x_{k,\ell_0} \right| \leq \sum_{k=r_p+1}^{\infty} |a_{m_p,n_p,k,\ell_0}|$$
$$< \frac{\epsilon}{8}, \quad \text{using (5.9);}$$

and

$$\left| \sum_{k=r_{p-1}+1}^{r_p} a_{m_p,n_p,k,\ell_0} x_{k,\ell_0} \right| = |a_{m_p,n_p,k_p,\ell_0}|$$
$$> \frac{3\epsilon}{4}, \quad \text{using (5.8).}$$

Thus, using the above inequalities,

$$
\begin{aligned}
|(Ax)_{m_p,n_p}| &= \left|\sum_{k,\ell=0}^{\infty,\infty} a_{m_p,n_p,k,\ell} x_{k,\ell}\right| \\
&= \left|\sum_{k=0}^{\infty} a_{m_p,n_p,k,\ell_0} x_{k,\ell_0}\right| \\
&\geq \left|\sum_{k=r_{p-1}+1}^{r_p} a_{m_p,n_p,k,\ell_0} x_{k,\ell_0}\right| \\
&\quad - \left|\sum_{k=0}^{r_{p-1}} a_{m_p,n_p,k,\ell_0} x_{k,\ell_0}\right| \\
&\quad - \left|\sum_{k=r_p+1}^{\infty} a_{m_p,n_p,k,\ell_0} x_{k,\ell_0}\right| \\
&> \frac{3\epsilon}{4} - \frac{\epsilon}{8} - \frac{\epsilon}{8} \\
&= \frac{\epsilon}{2}, \quad p = 0, 1, 2, \ldots .
\end{aligned}
$$

Thus, $\lim_{m+n\to\infty} (Ax)_{m,n} = 0$ does not hold, which is a contradiction. Thus, (5.4) is necessary. The necessity of (5.5) follows in a similar fashion. To prove (5.1), we shall suppose that (5.1) does not hold and arrive at a contradiction. Let $0 < \rho < 1$. Choose $m_0 = n_0 = 1$. Using (5.2) and (5.3), choose $m_1 + n_1 > m_0 + n_0$ such that

$$\sum_{k+\ell=0}^{m_0+n_0} |a_{m_1,n_1,k,\ell}| < 2, \quad \text{using (5.2);}$$

$$\sum_{k+\ell=0}^{\infty} |a_{m_1,n_1,k,\ell}| > \left(\frac{2}{\rho}\right)^6;$$

and

$$\sum_{k+\ell=m_0+n_0+1}^{\infty} |a_{m_1,n_1,k,\ell}| < 2^2$$

using (5.3), Theorems 5.1 and 5.2. It now follows that

$$\sum_{k+\ell=m_1+n_1+1}^{\infty} |a_{m_1,n_1,k,\ell}| < 2^2.$$

Choose $m_2 + n_2 > m_1 + n_1$ such that

$$\sum_{k+\ell=0}^{m_1+n_1} |a_{m_2,n_2,k,\ell}| < 2^2;$$

$$\sum_{k+\ell=0}^{\infty} |a_{m_2,n_2,k,\ell}| > \left(\frac{2}{\rho}\right)^8;$$

and

$$\sum_{k+\ell=m_2+n_2+1}^{\infty} |a_{m_2,n_2,k,\ell}| < 2^4.$$

Inductively, choose $m_p + n_p > m_{p-1} + n_{p-1}$ such that

$$\sum_{k+\ell=0}^{m_{p-1}+n_{p-1}} |a_{m_p,n_p,k,\ell}| < 2^{p-1}; \tag{5.10}$$

$$\sum_{k+\ell=0}^{\infty} |a_{m_p,n_p,k,\ell}| > \left(\frac{2}{\rho}\right)^{2p+2}; \tag{5.11}$$

and

$$\sum_{k+\ell=m_p+n_p+1}^{\infty} |a_{m_p,n_p,k,\ell}| < 2^{2p-2}. \tag{5.12}$$

Using (5.10)–(5.12), we have,

$$\begin{aligned}
&\sum_{k+\ell=m_{p-1}+n_{p-1}+1}^{m_p+n_p} |a_{m_p,n_p,k,\ell}| \\
&> \left(\frac{2}{\rho}\right)^{2p+2} - 2^{2p-2} - 2^{p-1} \\
&\geq \left(\frac{2}{\rho}\right)^{2p+2} - \left(\frac{2}{\rho}\right)^{2p-2} - \left(\frac{2}{\rho}\right)^{p-1}, \quad \text{since } \frac{1}{\rho} > 1 \\
&= \left(\frac{2}{\rho}\right)^{p-1} \left[\left(\frac{2}{\rho}\right)^{p+3} - \left(\frac{2}{\rho}\right)^{p-1} - 1\right] \\
&\geq \left(\frac{2}{\rho}\right)^{p-1} \left[\left(\frac{2}{\rho}\right)^{p+3} - \left(\frac{2}{\rho}\right)^{p-1} - \left(\frac{2}{\rho}\right)^{p-1}\right], \text{ since } \left(\frac{2}{\rho}\right)^{p-1} \geq 1
\end{aligned}$$

$$
\begin{aligned}
&= \left(\frac{2}{\rho}\right)^{p-1}\left[\left(\frac{2}{\rho}\right)^{4}\left(\frac{2}{\rho}\right)^{p-1} - 2\left(\frac{2}{\rho}\right)^{p-1}\right] \\
&> \left(\frac{2}{\rho}\right)^{p-1}\left[\left(\frac{2}{\rho}\right)^{4}\left(\frac{2}{\rho}\right)^{p-1} - \frac{2}{\rho}\left(\frac{2}{\rho}\right)^{p-1}\right], \text{ since } \frac{2}{\rho} > 2 \\
&= \left(\frac{2}{\rho}\right)^{2p-1}\left[\left(\frac{2}{\rho}\right)^{3} - 1\right] \\
&> \left(\frac{2}{\rho}\right)^{2p-1}[2^3 - 1], \text{ since } \frac{2}{\rho} > 2 \\
&= 7\left(\frac{2}{\rho}\right)^{2p-1} \\
&> 4\left(\frac{2}{\rho}\right)^{2p-1} \\
&= \frac{2^{2p+1}}{\rho^{2p-1}} \\
&> \frac{2^{2p+1}}{\rho^{p}}, \text{ since } \frac{1}{\rho} > 1.
\end{aligned}
$$

Thus, there exist $k_p, \ell_p, m_{p-1} + n_{p-1} < k_p + \ell_p \leq m_p + n_p$ such that

$$|a_{m_p,n_p,k_p,\ell_p}| > \frac{2^{2p+1}}{\rho^p}, \quad p = 0, 1, 2, \ldots. \tag{5.13}$$

Now, define the double sequence $\{x_{k,\ell}\}$ as follows:

$$x_{k,\ell} = \begin{cases} \rho^p sgn(a_{m_p,n_p,k_p,\ell_p}), & \text{if } k = k_p, \ell = \ell_p, p = 0, 1, 2, \ldots; \\ 0, & \text{otherwise.} \end{cases}$$

Note that $\lim_{k+\ell\to\infty} x_{k,\ell} = 0$. Now,

$$
\begin{aligned}
|(Ax)_{m_p,n_p}| &= \left|\sum_{k,\ell=0}^{\infty,\infty} a_{m_p,n_p,k,\ell} x_{k,\ell}\right| \\
&\geq \left|\sum_{k+\ell=m_{p-1}+n_{p-1}+1}^{m_p+n_p} a_{m_p,n_p,k,\ell} x_{k,\ell}\right| \\
&\quad - \left|\sum_{k+\ell=0}^{m_{p-1}+n_{p-1}} a_{m_p,n_p,k,\ell} x_{k,\ell}\right|
\end{aligned}
$$

$$
\begin{aligned}
&- \left| \sum_{k+\ell=m_p+n_p+1}^{\infty} a_{m_p,n_p,k,\ell} x_{k,\ell} \right| \\
&\geq \rho^p |a_{m_p,n_p,k_p,\ell_p}| - \sum_{k+\ell=0}^{m_{p-1}+n_{p-1}} |a_{m_p,n_p,k,\ell}| \\
&\quad - \sum_{k+\ell=m_p+n_p+1}^{\infty} |a_{m_p,n_p,k,\ell}| \\
&> \frac{2^{2p+1}}{\rho^p} \rho^p - 2^{2p-2} - 2^{p-1}, \text{ using (5.10), (5.12) and (5.13)} \\
&= 2^{2p+1} - 2^{2p-2} - 2^{p-1} \\
&= 2^{2p-2}(2^3 - 1) - 2^{p-1} \\
&= 2^{2p-2}(7) - 2^{p-1} \\
&= 2^{p-1}[7.2^{p-1} - 1] \\
&\geq 2^{p-1}[7.2^{p-1} - 2^{p-2}] \\
&= 2^{p-1}[2^{p-2}(14 - 1)] \\
&= 2^{p-1}[13.2^{p-2}] \\
&= 13.2^{2p-3},
\end{aligned}
$$

$$
i.e., \; |(Ax)_{m_p,n_p}| > 13.2^{2p-3}, \quad p = 0, 1, 2, \ldots,
$$

i.e., $\lim_{m+n\to\infty} (Ax)_{m,n} = 0$ does not hold, which is a contradiction. Thus, (5.1) is necessary too.

Proof of the sufficiency part. Let $\lim_{m+n\to\infty} x_{m,n} = x$. Then,

$$
(Ax)_{m,n} - x = \sum_{k,\ell=0}^{\infty,\infty} a_{m,n,k,\ell} x_{k,\ell} - x.
$$

Using (5.3), we have,

$$
\sum_{k,\ell=0}^{\infty,\infty} a_{m,n,k,\ell} - r_{m,n} = 1,
$$

where

$$
\lim_{m+n\to\infty} r_{m,n} = 0. \tag{5.14}
$$

Hence,

$$
(Ax)_{m,n} - x = \sum_{k,\ell=0}^{\infty,\infty} a_{m,n,k,\ell}(x_{k,\ell} - x) + r_{m,n} x.
$$

Using (5.1), we have,

$$|a_{m,n,k,\ell}| \leq H, \quad m, n, k, \ell = 0, 1, 2, \ldots, H > 0.$$

Given $\epsilon > 0$, we can choose sufficiently large p and q such that

$$\sum_{k+\ell=p+q+1}^{\infty} |x_{k,\ell} - x| < \frac{\epsilon}{5H}. \tag{5.15}$$

Let

$$|x_{k,\ell} - x| \leq L, \quad k + \ell = 0, 1, 2, \ldots, L > 0.$$

We now choose $N \in \mathbb{N}$ such that, whenever $m + n \geq N$, the following are satisfied:

$$\sum_{k,\ell=0}^{p,q} |a_{m,n,k,\ell}| < \frac{\epsilon}{5L}, \text{ using (5.2)}; \tag{5.16}$$

$$\sum_{k=0}^{\infty} |a_{m,n,k,\ell}| < \frac{\epsilon}{5L}, \ \ell = 0, 1, 2, \ldots, q, \text{ using (5.4)}; \tag{5.17}$$

$$\sum_{\ell=0}^{\infty} |a_{m,n,k,\ell}| < \frac{\epsilon}{5L}, \ k = 0, 1, 2, \ldots, p, \text{ using (5.5)}; \tag{5.18}$$

and

$$|r_{m,n}| < \frac{\epsilon}{5|x|}, \text{ using (5.14)}. \tag{5.19}$$

So, whenever $m + n \geq N$, we thus have,

$$\begin{aligned}
|(Ax)_{m,n} - x| &= \left| \sum_{k,\ell=0}^{\infty,\infty} a_{m,n,k,\ell}(x_{k,\ell} - x) + r_{m,n}x \right| \\
&\leq \left| \sum_{k,\ell=0}^{p,q} a_{m,n,k,\ell}(x_{k,\ell} - x) \right| \\
&\quad + \left| \sum_{k=0,\ell=q+1}^{p,\infty} a_{m,n,k,\ell}(x_{k,\ell} - x) \right|
\end{aligned}$$

$$
\begin{aligned}
&+ \left| \sum_{k=p+1,\ell=0}^{\infty,q} a_{m,n,k,\ell}(x_{k,\ell} - x) \right| \\
&+ \left| \sum_{k=p+1,\ell=q+1}^{\infty,\infty} a_{m,n,k,\ell}(x_{k,\ell} - x) \right| \\
&+ |r_{m,n}||x| \\
< &\frac{\epsilon}{5L} L + \frac{\epsilon}{5L} L + \frac{\epsilon}{5L} L + \frac{\epsilon}{5H} H + \frac{\epsilon}{5|x|}|x|, \\
&\text{using (5.15)–(5.19)} \\
= &\ \epsilon.
\end{aligned}
$$

Consequently,

$$\lim_{m+n\to\infty} (Ax)_{m,n} = x.$$

Thus, A is regular, completing the proof of the theorem. □

5.3 The Schur and Steinhaus Theorems for Four-Dimensional Infinite Matrices

Let c_{ds}, ℓ_{ds}^{∞}, respectively, denote the spaces of convergent double sequences and bounded double sequences.

Definition 5.5 A is called Schur matrix if $\{(Ax)_{m,n}\} \in c_{ds}$, whenever $x = \{x_{k,\ell}\} \in \ell_{ds}^{\infty}$.

In this section, we obtain necessary and sufficient conditions for $A = (a_{m,n,k,\ell})$ to be a Schur matrix and then deduce Steinhaus theorem.

Definition 5.6 The double sequence $\{x_{m,n}\}$ is called a Cauchy sequence if for every $\epsilon > 0$, there exists $N \in \mathbb{N}$ (the set of all positive integers) such that the set

$$\{(m,n), (k,\ell) \in \mathbb{N}^2 : |x_{m,n} - x_{k,\ell}| \geq \epsilon,\ m, n, k, \ell \geq N\}$$

is finite.

It is now easy to prove the following result.

Theorem 5.4 *The double sequence $\{x_{m,n}\}$ is Cauchy if and only if*

$$\lim_{m+n\to\infty} |x_{m+u,n} - x_{m,n}| = 0,\ u = 0, 1, 2, \ldots; \tag{5.20}$$

and

$$\lim_{m+n\to\infty} |x_{m,n+v} - x_{m,n}| = 0, \; v = 0, 1, 2, \ldots. \tag{5.21}$$

Definition 5.7 If every Cauchy double sequence of a normed linear space X converges to an element of X, X is said to be "double sequence complete" or "ds-complete".

Note that $\mathbb{R}$ (the set of all real numbers) and $\mathbb{C}$ (the set of all complex numbers) are ds-complete.

For $x = \{x_{m,n}\} \in \ell^{\infty}_{ds}$, define

$$\|x\| = \sup_{m,n\geq 0} |x_{m,n}|.$$

One can easily prove that ℓ^{∞}_{ds} is a normed linear space which is ds-complete. With the same definition of norm for elements of c_{ds}, c_{ds} is a closed subspace of ℓ^{∞}_{ds}.

Theorem 5.5 (Schur) *The necessary and sufficient conditions for a four-dimensional infinite matrix $A = (a_{m,n,k,\ell})$ to be a Schur matrix, i.e., $\{(Ax)_{m,n}\} \in c_{ds}$, whenever $x = \{x_{k,\ell}\} \in \ell^{\infty}_{ds}$ are as follows:*

$$\sum_{k,\ell=0}^{\infty,\infty} |a_{m,n,k,\ell}| < \infty, \; m, n = 0, 1, 2, \ldots; \tag{5.22}$$

$$\lim_{m+n\to\infty} \sum_{k,\ell=0}^{\infty,\infty} |a_{m+u,n,k,\ell} - a_{m,n,k,\ell}| = 0, \; u = 0, 1, 2, \ldots; \tag{5.23}$$

and

$$\lim_{m+n\to\infty} \sum_{k,\ell=0}^{\infty,\infty} |a_{m,n+v,k,\ell} - a_{m,n,k,\ell}| = 0, \; v = 0, 1, 2, \ldots. \tag{5.24}$$

Proof Proof of the sufficiency part. Let (5.22)–(5.24) hold and $x = \{x_{k,\ell}\} \in \ell^{\infty}_{ds}$. First, we note that in view of (5.22),

$$(Ax)_{m,n} = \sum_{k,\ell=0}^{\infty,\infty} a_{m,n,k,\ell} x_{k,\ell}, \; m, n = 0, 1, 2, \ldots$$

is defined, the double series on the right being convergent. Now, for $n = 0, 1, 2, \ldots,$

$$
\begin{aligned}
|(Ax)_{m+u,n} - (Ax)_{m,n}| &= \left| \sum_{k,\ell=0}^{\infty,\infty} (a_{m+u,n,k,\ell} - a_{m,n,k,\ell}) x_{k,\ell} \right| \\
&\leq M \sum_{k,\ell=0}^{\infty,\infty} |a_{m+u,n,k,\ell} - a_{m,n,k,\ell}| \\
&\to 0, \ m+n \to \infty, \text{ using (5.23)},
\end{aligned}
$$

where $|x_{k,\ell}| \leq M, k, \ell = 0, 1, 2, \ldots, M > 0$. Similarly, it follows that

$$|(Ax)_{m,n+v} - (Ax)_{m,n}| \to 0, \ m+n \to \infty, v = 0, 1, 2, \ldots,$$

using (5.24). Thus, $\{(Ax)_{m,n}\}$ is a Cauchy double sequence. Since $\mathbb{R}$ (or $\mathbb{C}$) is ds-complete, $\{(Ax)_{m,n}\}$ converges, i.e., $\{(Ax)_{m,n}\} \in c_{ds}$, completing the sufficiency part of the proof.

Proof of the necessity part. Let A be a Schur matrix. For $m, n = 0, 1, 2, \ldots$, consider the double sequence $\{x_{k,\ell}\}$, where $x_{k,\ell} = sgn(a_{m,n,k,\ell}), k, \ell = 0, 1, 2, \ldots$. Then, $\{x_{k,\ell}\} \in \ell_{ds}^{\infty}$ so that, by hypothesis,

$$(Ax)_{m,n} = \sum_{k,\ell=0}^{\infty,\infty} |a_{m,n,k,\ell}|, \ m, n = 0, 1, 2, \ldots$$

is defined. Since the series on the right converges, (5.22) holds. Suppose (5.23) does not hold. So, there exist $\ell_0, u_0 \in \mathbb{N}$ such that

$$\lim_{m+n\to\infty} \sum_{k=0}^{\infty} |a_{m+u_0,n,k,\ell_0} - a_{m,n,k,\ell_0}| = 0$$

does not hold. So, there exists $\epsilon > 0$ such that the set

$$\left\{ (m, n) \in \mathbb{N}^2 : \sum_{k=0}^{\infty} |a_{m+u_0,n,k,\ell_0} - a_{m,n,k,\ell_0}| > 2\epsilon \right\}$$

is infinite. Thus, we can choose pairs of integers $m_p, n_p \in \mathbb{N}$ such that $m_1 + n_1 < m_2 + n_2 < \cdots < m_p + n_p < \ldots$ and

$$\sum_{k=0}^{\infty} |a_{m_p+u_0,n_p,k,\ell_0} - a_{m_p,n_p,k,\ell_0}| > 2\epsilon, \ p = 1, 2, \ldots. \tag{5.25}$$

Using (5.22), we have,

$$\sum_{k=0}^{\infty} |a_{m_1+u_0,n_1,k,\ell_0} - a_{m_1,n_1,k,\ell_0}| < \infty.$$

Consequently, there exists $r_1 \in \mathbb{N}$ such that

$$\sum_{k=r_1}^{\infty} |a_{m_1+u_0,n_1,k,\ell_0} - a_{m_1,n_1,k,\ell_0}| < \frac{\epsilon}{4}. \tag{5.26}$$

In view of (5.25) and (5.26), we have,

$$\sum_{k=0}^{r_1-1} |a_{m_1+u_0,n_1,k,\ell_0} - a_{m_1,n_1,k,\ell_0}| > \frac{7\epsilon}{4} > \epsilon.$$

By hypothesis, (5.2) holds so that we can suppose that

$$\sum_{k=0}^{r_1-1} |a_{m_2+u_0,n_2,k,\ell_0} - a_{m_2,n_2,k,\ell_0}| < \frac{\epsilon}{4}. \tag{5.27}$$

Using (5.25), we have,

$$\sum_{k=0}^{\infty} |a_{m_2+u_0,n_2,k,\ell_0} - a_{m_2,n_2,k,\ell_0}| > 2\epsilon. \tag{5.28}$$

Using (5.1),

$$\sum_{k=0}^{\infty} |a_{m_2+u_0,n_2,k,\ell_0} - a_{m_2,n_2,k,\ell_0}| < \infty,$$

so that there exists $r_2 \in \mathbb{N}$, $r_2 > r_1$ such that

$$\sum_{k=r_2}^{\infty} |a_{m_2+u_0,n_2,k,\ell_0} - a_{m_2,n_2,k,\ell_0}| < \frac{\epsilon}{4}. \tag{5.29}$$

From (5.27)–(5.29), we have,

$$\sum_{k=r_1}^{r_2-1} |a_{m_2+u_0,n_2,k,\ell_0} - a_{m_2,n_2,k,\ell_0}| > \frac{3\epsilon}{2} > \epsilon.$$

Inductively, we can choose a strictly increasing sequence $\{r_p\}$ of positive integers such that

$$\sum_{k=0}^{r_{p-1}-1} |a_{m_p+u_0,n_p,k,\ell_0} - a_{m_p,n_p,k,\ell_0}| < \frac{\epsilon}{4}; \tag{5.30}$$

$$\sum_{k=r_p}^{\infty} |a_{m_p+u_0,n_p,k,\ell_0} - a_{m_p,n_p,k,\ell_0}| < \frac{\epsilon}{4}; \tag{5.31}$$

and

$$\sum_{k=r_{p-1}}^{r_p-1} |a_{m_p+u_0,n_p,k,\ell_0} - a_{m_p,n_p,k,\ell_0}| > \epsilon. \tag{5.32}$$

Now, define $\{x_{k,\ell}\} \in \ell_{ds}^{\infty}$, where

$$x_{k,\ell} = \begin{cases} sgn(a_{m_p+u_0,n_p,k,\ell_0} - a_{m_p,n_p,k,\ell_0}), & \text{if } \ell = \ell_0, r_{p-1} \leq k < r_p, p = 1, 2, \ldots; \\ 0, & \text{otherwise.} \end{cases}$$

Then,

$$\begin{aligned}
(Ax)_{m_p+u_0,n_p} - (Ax)_{m_p,n_p} &= \sum_{k,\ell=0}^{\infty,\infty} (a_{m_p+u_0,n_p,k,\ell} - a_{m_p,n_p,k,\ell})x_{k,\ell} \\
&= \sum_{k=0}^{\infty} (a_{m_p+u_0,n_p,k,\ell_0} - a_{m_p,n_p,k,\ell_0})x_{k,\ell_0} \\
&= \sum_{k=0}^{r_{p-1}-1} (a_{m_p+u_0,n_p,k,\ell_0} - a_{m_p,n_p,k,\ell_0})x_{k,\ell_0} \\
&\quad + \sum_{k=r_{p-1}}^{r_p-1} (a_{m_p+u_0,n_p,k,\ell_0} - a_{m_p,n_p,k,\ell_0})x_{k,\ell_0} \\
&\quad + \sum_{k=r_p}^{\infty} (a_{m_p+u_0,n_p,k,\ell_0} - a_{m_p,n_p,k,\ell_0})x_{k,\ell_0} \\
&= \sum_{k=0}^{r_{p-1}-1} (a_{m_p+u_0,n_p,k,\ell_0} - a_{m_p,n_p,k,\ell_0})x_{k,\ell_0} \\
&\quad + \sum_{k=r_{p-1}}^{r_p-1} |a_{m_p+u_0,n_p,k,\ell_0} - a_{m_p,n_p,k,\ell_0}| \\
&\quad + \sum_{k=r_p}^{\infty} (a_{m_p+u_0,n_p,k,\ell_0} - a_{m_p,n_p,k,\ell_0})x_{k,\ell_0}
\end{aligned}$$

so that

$$\begin{aligned}\sum_{k=r_{p-1}}^{r_p-1} |a_{m_p+u_0,n_p,k,\ell_0} - a_{m_p,n_p,k,\ell_0}| &= \{(Ax)_{m_p+u_0,n_p} - (Ax)_{m_p,n_p}\} \\ &\quad - \sum_{k=0}^{r_{p-1}-1} (a_{m_p+u_0,n_p,k,\ell_0} - a_{m_p,n_p,k,\ell_0})x_{k,\ell_0} \\ &\quad - \sum_{k=r_p}^{\infty}(a_{m_p+u_0,n_p,k,\ell_0} - a_{m_p,n_p,k,\ell_0})x_{k,\ell_0}.\end{aligned}$$

In view of (5.30)–(5.32), we have,

$$\begin{aligned}\epsilon &< \sum_{k=r_{p-1}}^{r_p-1} |a_{m_p+u_0,n_p,k,\ell_0} - a_{m_p,n_p,k,\ell_0}| \\ &\leq \left|(Ax)_{m_p+u_0,n_p} - (Ax)_{m_p,n_p}\right| + \frac{\epsilon}{4} + \frac{\epsilon}{4},\end{aligned}$$

from which it follows that

$$\left|(Ax)_{m_p+u_0,n_p} - (Ax)_{m_p,n_p}\right| > \frac{\epsilon}{2},\ p = 1, 2, \ldots.$$

Consequently,

$$\{(Ax)_{m,n}\} \notin c_{ds},$$

which is a contradiction. Thus, (5.23) holds. Similarly, (5.24) holds too. This completes the proof of the theorem. □

We now deduce the following result.

Theorem 5.6 (Steinhaus) *A four-dimensional infinite matrix $A = (a_{m,n,k,\ell})$ cannot be both a regular and a Schur matrix, i.e., given a four-dimensional regular matrix A, there exists a bounded, divergent double sequence which is not A-summable.*

Proof Since A is regular, (5.2) and (5.3) hold. If A were a Schur matrix too, then $\{a_{m,n,k,\ell}\}_{m,n=0}^{\infty,\infty}$ is uniformly Cauchy with respect to $k, \ell = 0, 1, 2, \ldots$. Since $\mathbb{R}$ (or $\mathbb{C}$) is ds-complete, $\{a_{m,n,k,\ell}\}_{m,n=0}^{\infty,\infty}$ converges uniformly to 0 with respect to k, $\ell = 0, 1, 2, \ldots$. Consequently, we have,

$$\lim_{m+n\to\infty} \sum_{k,\ell=0}^{\infty,\infty} a_{m,n,k,\ell} = 0,$$

a contradiction of (5.3), completing the proof. □

References

1. Natarajan, P.N., Srinivasan, V.: Silverman-Toeplitz theorem for double sequences and series and its application to Nörlund means in non-archimedean fields. Ann. Math. Blaise Pascal **9**, 85–100 (2002)
2. Kojima, T.: On the theory of double sequences. Tôhoku Math. J. **21**, 3–14 (1922)
3. Robison, G.M.: Divergent double sequences and series. Trans. Amer. Math. Soc. **28**, 50–73 (1926)
4. Natarajan, P.N.: A new definition of convergence of a double sequence and a double series and Silverman-Toeplitz theorem. Comment. Math. Prace Mat. **54**, 129–139 (2014)
5. Natarajan, P.N.: The Schur and Steinhaus theorems for 4-dimensional infinite matrices. Comment. Math. Prace Mat. **54**, 159–165 (2014)

Chapter 6
The Nörlund, Weighted Mean, and $(M, \lambda_{m,n})$ Methods for Double Sequences

The present chapter is devoted to a study of the Nörlund, Weighted Mean, and $(M, \lambda_{m,n})$ or Natarajan methods for double sequences. This chapter is divided into 3 sections. In the first section, the Nörlund method for double sequences is introduced and its properties are studied in the context of the new definition of convergence of a double sequence introduced earlier in Chap. 5. In the second section, we introduce the Weighted Mean method for double sequences and study its properties in the context of the new definition. In the final section, the $(M, \lambda_{m,n})$ method or the Natarajan method for double sequences is introduced and its properties are studied in the context of the new definition.

6.1 Nörlund Method for Double Sequences

We now introduce Nörlund methods for double sequences and double series and study them in the context of Definitions 5.1 and 5.2 (see [1]).

Definition 6.1 Given a doubly infinite set of real numbers $p_{m,n}$, $m, n = 0, 1, 2, \ldots$, where

$$p_{m,n} \geq 0, \ (m, n) \neq (0, 0) \text{ and } p_{0,0} > 0,$$

let

$$P_{m,n} = \sum_{i,j=0}^{m,n} p_{i,j}, \quad m, n = 0, 1, 2, \ldots.$$

For any double sequence $\{s_{m,n}\}$, we define

P.N. Natarajan, *Classical Summability Theory*, DOI 10.1007/978-981-10-4205-8_6

$$\begin{aligned}\sigma_{m,n} &= (N, p_{m,n})(\{s_{m,n}\})\\ &= \frac{S_{m,n}}{P_{m,n}}\\ &= \frac{1}{P_{m,n}}\left[\sum_{i,j=0}^{m,n} p_{m-i,n-j}s_{i,j}\right], \quad m, n = 0, 1, 2, \ldots.\end{aligned}$$

If $\lim_{m+n\to\infty} \sigma_{m,n} = \sigma$, we say that the double sequence $\{s_{m,n}\}$ is summable $(N, p_{m,n})$ to σ, written as

$$s_{m,n} \to \sigma(N, p_{m,n}).$$

$(N, p_{m,n})$ is called a Nörlund method (or Nörlund mean).

Definition 6.2 A double series $\sum_{m,n=0}^{\infty,\infty} u_{m,n}$ is said to be $(N, p_{m,n})$ summable to σ, if the double sequence $\{s_{m,n}\}$, where

$$s_{m,n} = \sum_{i,j=0}^{m,n} u_{i,j}, \quad m, n = 0, 1, 2, \ldots,$$

is $(N, p_{m,n})$ summable to σ.

Definition 6.3 The Nörlund methods $(N, p_{m,n})$ and $(N, q_{m,n})$ are said to be consistent if

$$s_{m,n} \to \sigma(N, p_{m,n})$$

and

$$s_{m,n} \to \sigma'(N, q_{m,n})$$

imply that $\sigma = \sigma'$.

Definition 6.4 We say that $(N, p_{m,n})$ is included in $(N, q_{m,n})$ (or $(N, q_{m,n})$ includes $(N, p_{m,n})$), written as

$$(N, p_{m,n}) \subseteq (N, q_{m,n})$$

if

$$s_{m,n} \to \sigma(N, p_{m,n}) \text{ implies that } s_{m,n} \to \sigma(N, q_{m,n}).$$

The two Nörlund methods $(N, p_{m,n})$, $(N, q_{m,n})$ are said to be equivalent if

$$(N, p_{m,n}) \subseteq (N, q_{m,n}) \text{ and vice versa.}$$

In view of Theorem 5.3, it is easy to prove the following result.

Theorem 6.1 *The Nörlund method $(N, p_{m,n})$ is regular if and only if*

$$\lim_{m+n\to\infty} \sum_{j=0}^{n} p_{m-i,n-j} = 0, \quad i = 0, 1, 2, \ldots, m; \tag{6.1}$$

and

$$\lim_{m+n\to\infty} \sum_{i=0}^{m} p_{m-i,n-j} = 0, \quad j = 0, 1, 2, \ldots, n. \tag{6.2}$$

In the sequel, let $(N, p_{m,n})$, $(N, q_{m,n})$ be regular Nörlund methods such that each row and each column of the two-dimensional infinite matrices $(p_{m,n})$, $(q_{m,n})$ is a regular Nörlund method for simple sequences.

Theorem 6.2 *Any two such regular Nörlund methods are consistent.*

Proof Given regular Nörlund methods $(N, p_{m,n})$, $(N, q_{m,n})$, where each row and each column of the two-dimensional infinite matrices $(p_{m,n})$, $(q_{m,n})$ is a regular Nörlund method for simple sequences, we define a third method $(N, r_{m,n})$ by the equation

$$r_{m,n} = \sum_{i,j=0}^{m,n} p_{i,j} q_{m-i,n-j}, \quad m, n = 0, 1, 2, \ldots.$$

Then, for $s = \{s_{m,n}\}$, we can check that

$$(N, r_{m,n})(s) = \sum_{i,j=0}^{\infty,\infty} \gamma_{m,n,i,j} (N, q_{i,j})(s),$$

where

$$\gamma_{m,n,i,j} = \begin{cases} \dfrac{p_{m-i,n-j} Q_{i,j}}{\sum\limits_{\mu,\gamma=0}^{m,n} p_{m-\mu,n-\gamma} Q_{\mu,\gamma}}, & \text{if } 0 \leq i \leq m \text{ and } 0 \leq j \leq n; \\ 0, & \text{otherwise.} \end{cases}$$

Using the fact that $p_{0,0} > 0$, $q_{0,0} > 0$, $p_{m,n} \geq 0$, $q_{m,n} \geq 0$, $(m, n) \neq (0, 0)$ and the fact that each row and each column of the two-dimensional infinite matrices $(p_{m,n})$, $(q_{m,n})$ is a regular Nörlund method for simple sequences, it follows that

$$\lim_{m+n\to\infty} \sum_{j=0}^{n} \gamma_{m,n,i,j} = 0, \quad i = 0, 1, 2, \ldots, m$$

and

$$\lim_{m+n\to\infty} \sum_{i=0}^{m} \gamma_{m,n,i,j} = 0, \quad j = 0, 1, 2, \ldots, n.$$

Appealing to Theorem 5.3, the four-dimensional infinite matrix $(\gamma_{m,n,k,\ell})$ is regular. Consequently,

$$s_{m,n} \to \sigma'(N, q_{m,n}) \text{ implies that } s_{m,n} \to \sigma'(N, r_{m,n}).$$

Similarly, it can be shown that

$$s_{m,n} \to \sigma(N, p_{m,n}) \text{ implies that } s_{m,n} \to \sigma(N, r_{m,n}).$$

So $\sigma = \sigma'$. Thus, the regular Nörlund methods $(N, p_{m,n})$ and $(N, q_{m,n})$ are consistent, completing the proof of the theorem. □

Let $(N, p_{m,n})$, $(N, q_{m,n})$ be regular Nörlund methods. Then,

$$P(x, y) = \sum P_{m,n} x^m y^n,$$
$$Q(x, y) = \sum Q_{m,n} x^m y^n,$$
$$p(x, y) = \sum p_{m,n} x^m y^n,$$
$$q(x, y) = \sum q_{m,n} x^m y^n$$

all converge for $|x|, |y| < 1$. Also,

$$k(x, y) = \sum k_{m,n} x^m y^n = \frac{q(x, y)}{p(x, y)} = \frac{Q(x, y)}{P(x, y)},$$
$$\ell(x, y) = \sum \ell_{m,n} x^m y^n = \frac{p(x, y)}{q(x, y)} = \frac{P(x, y)}{Q(x, y)}$$

converge for $|x|, |y| < 1$.

Further, we have

$$\sum_{i,j=0}^{m,n} k_{i,j} p_{m-i,n-j} = q_{m,n};$$
$$\sum_{i,j=0}^{m,n} k_{i,j} P_{m-i,n-j} = Q_{m,n};$$
$$\sum_{i,j=0}^{m,n} \ell_{i,j} q_{m-i,n-j} = p_{m,n};$$

and

$$\sum_{i,j=0}^{m,n} \ell_{i,j} Q_{m-i,n-j} = P_{m,n}.$$

We now have

Theorem 6.3 *Let* $(N, p_{m,n})$, $(N, q_{m,n})$ *be regular Nörlund methods. Then*

$$(N, p_{m,n}) \subseteq (N, q_{m,n})$$

if and only if

$$\frac{\sum_{i,j=0}^{m,n} |k_{i,j}| P_{m-i,n-j}}{Q_{m,n}} = O(1), \quad m + n \to \infty,$$

i.e.,

$$\sum_{i,j=0}^{m,n} |k_{i,j}| P_{m-i,n-j} \leq H Q_{m,n}, \quad m, n = 0, 1, 2, \ldots,$$

$$H > 0 \text{ is independent of } m, n; \tag{6.3}$$

$$\lim_{m+n\to\infty} \frac{\sum_{j=0}^{n} |k_{i,j}| P_{m-i,n-j}}{Q_{m,n}} = 0, \quad 0 \leq i \leq m; \tag{6.4}$$

and

$$\lim_{m+n\to\infty} \frac{\sum_{i=0}^{m} |k_{i,j}| P_{m-i,n-j}}{Q_{m,n}} = 0, \quad 0 \leq j \leq n. \tag{6.5}$$

Proof Let

$$s(x, y) = \sum s_{m,n} x^m y^n.$$

Then, for small x and y,

$$\sum Q_{m,n} N_{m,n}^{(q)}(s) x^m y^n = \sum \left(\sum_{i,j=0}^{m,n} q_{m-i,n-j} s_{i,j} \right) x^m y^n$$
$$= q(x, y) s(x, y);$$

Similarly,

$$\sum P_{m,n} N_{m,n}^{(p)}(s) x^m y^n = p(x, y) s(x, y).$$

Thus,

$$\frac{\sum Q_{m,n} N_{m,n}^{(q)}(s) x^m y^n}{\sum P_{m,n} N_{m,n}^{(p)}(s) x^m y^n} = \frac{q(x, y)}{p(x, y)}$$
$$= k(x, y),$$

$$i.e., \sum Q_{m,n} N_{m,n}^{(q)}(s) x^m y^n = k(x, y) \sum P_{m,n} N_{m,n}^{(p)}(s) x^m y^n$$
$$= \left(\sum k_{m,n} x^m y^n \right) \left(\sum P_{m,n} N_{m,n}^{(p)}(s) x^m y^n \right)$$

and so

$$Q_{m,n} N_{m,n}^{(q)}(s) = \sum_{i,j=0}^{m,n} k_{m-i,n-j} P_{i,j} N_{i,j}^{(p)}(s), \quad m, n = 0, 1, 2, \ldots.$$

Consequently,

$$N_{m,n}^{(q)}(s) = \sum_{i,j=0}^{\infty,\infty} c_{m,n,i,j} N_{i,j}^{(p)}(s),$$

where the four-dimensional infinite matrix $(c_{m,n,i,j})$ is defined by

$$c_{m,n,i,j} = \begin{cases} \frac{k_{m-i,n-j}}{Q_{m,n}}, & \text{if } 0 \leq i \leq m \text{ and } 0 \leq j \leq n; \\ 0, & \text{otherwise.} \end{cases}$$

Now,

$$(N, p_{m,n}) \subseteq (N, q_{m,n})$$

if and only if the four-dimensional infinite matrix $(c_{m,n,i,j})$ is regular. Rest of the proof follows using Theorem 5.3. This completes the proof of the theorem. □

Theorem 6.4 *The regular Nörlund methods* $(N, p_{m,n})$, $(N, q_{m,n})$ *are equivalent if and only if*

$$\sum_{m,n=0}^{\infty,\infty} |k_{m,n}| < \infty \text{ and } \sum_{m,n=0}^{\infty,\infty} |\ell_{m,n}| < \infty. \tag{6.6}$$

Proof Necessity part. Let $(N, p_{m,n})$, $(N, q_{m,n})$ be equivalent. Note that $k_{0,0} > 0$ and $\ell_{0,0} > 0$, since $p_{0,0} > 0$ and $q_{0.0} > 0$. Since $(N, p_{m,n}) \subseteq (N, q_{m,n})$, in view of (6.3),

$$k_{0,0} P_{m,n} \leq H Q_{m,n}, \quad m, n = 0, 1, 2, \ldots,$$

so that $\left\{\frac{P_{m,n}}{Q_{m,n}}\right\}$ is bounded. Similarly, $(N, q_{m,n}) \subseteq (N, p_{m,n})$ implies that $\left\{\frac{Q_{m,n}}{P_{m,n}}\right\}$ is bounded too. Using (6.3) again, we have

$$\sum_{i,j=0}^{\mu,\gamma} |k_{i,j}| \leq HL,$$

where $\left|\frac{Q_{m,n}}{P_{m,n}}\right| \leq L$, $m, n = 0, 1, 2, \ldots$. Thus, $\sum_{m,n=0}^{\infty,\infty} |k_{m,n}| < \infty$. Similarly, $\sum_{m,n=0}^{\infty,\infty} |\ell_{m,n}| < \infty$.

Sufficiency part. Let (6.6) hold. $\sum_{m,n=0}^{\infty,\infty} |k_{m,n}| < \infty$ and $\sum_{m,n=0}^{\infty,\infty} |\ell_{m,n}| < \infty$ in the sense of Definition 5.2 implies $\sum_{m,n=0}^{\infty,\infty} |k_{m,n}| < \infty$ and $\sum_{m,n=0}^{\infty,\infty} |\ell_{m,n}| < \infty$ in Pringsheim's sense, in view of Theorem 5.1. It now follows that the regular Nörlund methods $(N, p_{m,n})$ and $(N, q_{m,n})$ are equivalent, using the sufficiency part of Theorem III of [2]. The proof of the theorem is now complete. □

6.2 Weighted Mean Method for Double Sequences

We introduce Weighted Mean method for double sequences and double series and study them in the context of the new definition of convergence of a double sequence and a double series introduced earlier (see Definitions 5.1 and 5.2). For details about this study, one can refer to [3, 4].

Definition 6.5 Given a doubly infinite set of real number $p_{m,n}$, $m, n = 0, 1, 2, \ldots$, where $p_{m,n} \geq 0$, $(m, n) \neq (0, 0)$ and $p_{0,0} > 0$, let

$$P_{m,n} = \sum_{i,j=0}^{m,n} p_{i,j}, \quad m, n = 0, 1, 2, \ldots.$$

For any double sequence $\{s_{m,n}\}$, define

$$\begin{aligned}\sigma_{m,n} &= (\overline{N}, p_{m,n})(\{s_{m,n}\}) \\ &= \frac{S_{m,n}}{P_{m,n}} \\ &= \frac{1}{P_{m,n}} \left[\sum_{i,j=0}^{m,n} p_{i,j} s_{i,j} \right], \quad m, n = 0, 1, 2, \ldots.\end{aligned}$$

If $\lim_{m+n\to\infty} \sigma_{m,n} = \sigma$, we say that the double sequence $\{s_{m,n}\}$ is summable $(\overline{N}, p_{m,n})$ or $(\overline{N}, p_{m,n})$-summable to σ, written as

$$s_{m,n} \to \sigma(\overline{N}, p_{m,n}).$$

$(\overline{N}, p_{m,n})$ is called a "Weighted Mean method."

Note that the $(\overline{N}, p_{m,n})$ method is defined by the four-dimensional infinite matrix $A = (a_{m,n,k,\ell})$, where

$$a_{m,n,k,\ell} = \begin{cases} \frac{p_{k,\ell}}{P_{m,n}}, & \text{if } 0 \le k \le m \text{ and } 0 \le \ell \le n; \\ 0, & \text{otherwise}, \end{cases}$$

where $p_{m,n} \geq 0$, $(m, n) \neq (0, 0)$, $p_{0,0} > 0$ and

$$P_{m,n} = \sum_{i,j=0}^{m,n} p_{i,j}, \quad m, n = 0, 1, 2, \ldots.$$

Definition 6.6 A double series $\sum_{m,n=0}^{\infty,\infty} u_{m,n}$ is said to be $(\overline{N}, p_{m,n})$-summable to σ if the double sequence $\{s_{m,n}\}$, where

$$s_{m,n} = \sum_{i,j=0}^{m,n} u_{i,j}, \quad m, n = 0, 1, 2, \ldots,$$

is $(\overline{N}, p_{m,n})$-summable to σ.

Definition 6.7 We say that $(\overline{N}, p_{m,n})$ is included in $(\overline{N}, q_{m,n})$ (or $(\overline{N}, q_{m,n})$ includes $(\overline{N}, p_{m,n})$), written as

$$(\overline{N}, p_{m,n}) \subseteq (\overline{N}, q_{m,n}) \ \text{(or } (\overline{N}, q_{m,n}) \supseteq (\overline{N}, p_{m,n}))$$

if

$$s_{m,n} \to \sigma(\overline{N}, p_{m,n}) \text{ implies that } s_{m,n} \to \sigma(\overline{N}, q_{m,n}).$$

In view of Theorem 5.3, we have the following result.

Theorem 6.5 *The Weighted Mean method $(\overline{N}, p_{m,n})$ is regular if and only if*

$$\lim_{m+n\to\infty} P_{m,n} = \infty; \tag{6.7}$$

$$\lim_{m+n\to\infty} \frac{\sum_{k=0}^{m} p_{k,\ell}}{P_{m,n}} = 0, \ \ell = 0, 1, 2, \ldots; \tag{6.8}$$

and

$$\lim_{m+n\to\infty} \frac{\sum_{\ell=0}^{n} p_{k,\ell}}{P_{m,n}} = 0, \ k = 0, 1, 2, \ldots. \tag{6.9}$$

Proof Proof of the necessity part. Let $(\overline{N}, p_{m,n})$ be regular. Using (5.2),

$$\lim_{m+n\to\infty} a_{m,n,0,0} = 0,$$
$$i.e., \ \lim_{m+n\to\infty} \frac{p_{0,0}}{P_{m,n}} = 0,$$
$$i.e., \ \lim_{m+n\to\infty} P_{m,n} = \infty,$$

since $p_{0,0} \neq 0$.

Using (5.4),

$$\lim_{m+n\to\infty} \sum_{k=0}^{\infty} |a_{m,n,k,\ell}| = 0, \ \ell = 0, 1, 2, \ldots,$$
$$i.e., \ \lim_{m+n\to\infty} \sum_{k=0}^{m} \frac{p_{k,\ell}}{P_{m,n}} = 0, \ \ell = 0, 1, 2, \ldots,$$
$$i.e., \ \lim_{m+n\to\infty} \frac{\sum_{k=0}^{m} p_{k,\ell}}{P_{m,n}} = 0, \ \ell = 0, 1, 2, \ldots,$$

proving (6.8). Proof of (6.9) is similar.

Proof of the sufficiency part. Let (6.7)–(6.9) hold. For every fixed $k, \ell = 0, 1, 2, \ldots,$

$$\lim_{m+n\to\infty} a_{m,n,k,\ell} = \lim_{m+n\to\infty} \frac{p_{k,\ell}}{P_{m,n}} = 0,$$

in view of (6.7) so that (5.2) holds. Now,

$$\lim_{m+n\to\infty} \sum_{k,\ell=0}^{\infty,\infty} a_{m,n,k,\ell} = \lim_{m+n\to\infty} \frac{\sum_{k,\ell=0}^{m,n} p_{k,\ell}}{P_{m,n}} = \lim_{m+n\to\infty} \frac{P_{m,n}}{P_{m,n}} = 1,$$

proving (5.3). Also, for every fixed $\ell = 0, 1, 2, \ldots,$

$$\lim_{m+n\to\infty} \sum_{k=0}^{\infty} |a_{m,n,k,\ell}| = \lim_{m+n\to\infty} \sum_{k=0}^{m} |a_{m,n,k,\ell}| = \lim_{m+n\to\infty} \frac{\sum_{k=0}^{m} p_{k,\ell}}{P_{m,n}} = 0, \text{ using (6.8)}.$$

This proves (5.4). Proof of (5.5) is similar. Now,

$$\sum_{k,\ell=0}^{\infty,\infty} |a_{m,n,k,\ell}| = \frac{\sum_{k,\ell=0}^{m,n} p_{k,\ell}}{P_{m,n}} = \frac{P_{m,n}}{P_{m,n}} = 1, \; m, n = 0, 1, 2, \ldots,$$

$$\sup_{m,n\geq 0} \sum_{k,\ell=0}^{\infty,\infty} |a_{m,n,k,\ell}| < \infty.$$

Thus, (5.1) holds. So, by Theorem 5.3, $(\overline{N}, p_{m,n})$ is regular, completing the proof of the theorem. □

Theorem 6.6 (Limitation theorem) *If* $\{s_{m,n}\}$ *is* $(\overline{N}, p_{m,n})$*-summable to* s*, then*

$$|s_{m,n} - s| = o\left(\frac{P_{m,n}}{p_{m,n}}\right), \quad m+n \to \infty, \tag{6.10}$$

in the sense that

$$\frac{p_{m,n}}{P_{m,n}}(s_{m,n} - s) \to 0, \quad m+n \to \infty.$$

Proof Let $\{t_{m,n}\}$ be the $(\overline{N}, p_{m,n})$-transform of $\{s_{m,n}\}$. Now,

$$\begin{aligned}
\left|\frac{p_{m,n}}{P_{m,n}}(s_{m,n} - s)\right| &= \left|\frac{p_{m,n}s_{m,n} - p_{m,n}s}{P_{m,n}}\right| \\
&= \left|\frac{\begin{array}{l} P_{m,n}t_{m,n} - P_{m,n-1}t_{m,n-1} - P_{m-1,n}t_{m-1,n} + P_{m-1,n-1}t_{m-1,n-1} \\ -(P_{m,n} - P_{m,n-1} - P_{m-1,n} + P_{m-1,n-1})s \end{array}}{P_{m,n}}\right| \\
&= \left|(t_{m,n} - s) - \frac{P_{m,n-1}}{P_{m,n}}(t_{m,n-1} - s) - \frac{P_{m-1,n}}{P_{m,n}}(t_{m-1,n} - s)\right. \\
&\qquad \left. + \frac{P_{m-1,n-1}}{P_{m,n}}(t_{m-1,n-1} - s)\right| \\
&\le |t_{m,n} - s| + |t_{m,n-1} - s| + |t_{m-1,n} - s| + |t_{m-1,n-1} - s|, \\
&\qquad \text{since } \frac{P_{m,n-1}}{P_{m,n}}, \frac{P_{m-1,n}}{P_{m,n}}, \frac{P_{m-1,n-1}}{P_{m,n}} \le 1 \\
&\to 0, \ m+n \to \infty, \text{ since } \lim_{m+n\to\infty} t_{m,n} = s.
\end{aligned}$$

So,

$$\lim_{m+n\to\infty}\left|\frac{p_{m,n}}{P_{m,n}}(s_{m,n} - s)\right| = 0,$$

$$i.e., \ |s_{m,n} - s| = o\left(\frac{P_{m,n}}{p_{m,n}}\right), \quad m+n \to \infty,$$

completing the proof of the theorem. □

We now prove a few inclusion theorems involving Weighted Mean methods for double sequences.

Theorem 6.7 (Comparison theorem for two Weighted Mean methods for double sequences) *If* $(\overline{N}, p_{m,n})$*,* $(\overline{N}, q_{m,n})$ *are two Weighted Mean methods such that*

$$\sum_{m,n=0}^{\infty,\infty} \frac{q_{m,n}}{p_{m,n}} < \infty; \tag{6.11}$$

and

$$P_{m,n} = o(Q_{m,n}),\ m+n \to \infty, \tag{6.12}$$

in the sense that

$$\frac{P_{m,n}}{Q_{m,n}} \to 0,\ m+n \to \infty,$$

where

$$P_{m,n} = \sum_{k,\ell=0}^{m,n} p_{k,\ell},\quad Q_{m,n} = \sum_{k,\ell=0}^{m,n} q_{k,\ell},\ m, n = 0, 1, 2, \ldots,$$

then,

$$(\overline{N}, p_{m,n}) \subseteq (\overline{N}, q_{m,n}).$$

Proof Given a double sequence $\{s_{m,n}\}$, let

$$t_{m,n} = \frac{\sum_{i,j=0}^{m,n} p_{i,j}s_{i,j}}{P_{m,n}};$$

$$u_{m,n} = \frac{\sum_{i,j=0}^{m,n} q_{i,j}s_{i,j}}{Q_{m,n}},\ m, n = 0, 1, 2, \ldots.$$

Then,

$$p_{0,0}s_{0,0} = P_{0,0}t_{0,0};$$

$$p_{m,n}s_{m,n} = P_{m,n}t_{m,n} - P_{m-1,n}t_{m-1,n} - P_{m,n-1}t_{m,n-1} + P_{m-1,n-1}t_{m-1,n-1},$$

so that

$$s_{m,n} = \frac{P_{m,n}t_{m,n} - P_{m-1,n}t_{m-1,n} - P_{m,n-1}t_{m,n-1} + P_{m-1,n-1}t_{m-1,n-1}}{p_{m,n}},$$

$$s_{m,0} = \frac{P_{m,0}t_{m,0} - P_{m-1,0}t_{m-1,0}}{p_{m,0}},$$

$$s_{0,n} = \frac{P_{0,n}t_{0,n} - P_{0,n-1}t_{0,n-1}}{p_{0,n}},$$

where we suppose that

$$P_{-1,n} = 0,\ P_{m,-1} = 0,\ P_{-1,-1} = 0.$$

Now,

$$\begin{aligned} u_{m,n} &= \frac{1}{Q_{m,n}} \left(\sum_{i,j=0}^{m,n} q_{i,j} s_{i,j} \right) \\ &= \frac{1}{Q_{m,n}} \sum_{i,j=0}^{m,n} q_{i,j} \left\{ \frac{\begin{array}{c} P_{i,j}t_{i,j} - P_{i-1,j}t_{i-1,j} - P_{i,j-1}t_{i,j-1} \\ +P_{i-1,j-1}t_{i-1,j-1} \end{array}}{p_{i,j}} \right\} \\ &= \sum_{k,\ell=0}^{\infty,\infty} c_{m,n,k,\ell} t_{k,\ell}, \end{aligned}$$

where

$$c_{m,n,k,\ell} = \begin{cases} \left(\frac{q_{k,\ell}}{p_{k,\ell}} - \frac{q_{k,\ell+1}}{p_{k,\ell+1}} - \frac{q_{k+1,\ell}}{p_{k+1,\ell}} + \frac{q_{k+1,\ell+1}}{p_{k+1,\ell+1}} \right) \frac{P_{k,\ell}}{Q_{m,n}}, & \text{if } k < m,\ \ell < n; \\ \left(\frac{q_{k,\ell}}{p_{k,\ell}} - \frac{q_{k,\ell+1}}{p_{k,\ell+1}} \right) \frac{P_{k,\ell}}{Q_{m,n}}, & \text{if } k = m,\ \ell < n; \\ \left(\frac{q_{k,\ell}}{p_{k,\ell}} - \frac{q_{k+1,\ell}}{p_{k+1,\ell}} \right) \frac{P_{k,\ell}}{Q_{m,n}}, & \text{if } k < m,\ \ell = n; \\ \frac{q_{k,\ell}}{p_{k,\ell}} \frac{P_{k,\ell}}{Q_{k,\ell}}, & \text{if } k = m,\ \ell = n; \\ 0, & \text{if } k > m \text{ or } \ell > n. \end{cases}$$

In view of (6.11) and (6.12), we have

$$\lim_{m+n\to\infty} c_{m,n,k,\ell} = 0, \quad k, \ell = 0, 1, 2, \ldots.$$

If $s_{m,n} = 1$, $m, n = 0, 1, 2, \ldots$, then

$$t_{m,n} = \frac{\sum_{i,j=0}^{m,n} p_{i,j}}{P_{m,n}} = 1,$$

noting that $P_{m,n} \neq 0$, $m, n = 0, 1, 2, \ldots$. Similarly, $u_{m,n} = 1$, $m, n = 0, 1, 2, \ldots$, so that

$$u_{m,n} = \sum_{k,\ell=0}^{\infty,\infty} c_{m,n,k,\ell} t_{k,\ell}$$

yields

$$1 = \sum_{k,\ell=0}^{\infty,\infty} c_{m,n,k,\ell}(1).$$

Thus,

$$\sum_{k,\ell=0}^{\infty,\infty} c_{m,n,k,\ell} = 1, \quad m, n = 0, 1, 2, \ldots$$

and consequently,

$$\lim_{m+n\to\infty} \sum_{k,\ell=0}^{\infty,\infty} c_{m,n,k,\ell} = 1.$$

$$\begin{aligned}
\sum_{k,\ell=0}^{\infty,\infty} |c_{m,n,k,\ell}| &= \sum_{k,\ell=0}^{m,n} |c_{m,n,k,\ell}| \\
&= \sum_{k,\ell=0}^{m-1,n-1} \left| \frac{q_{k,\ell}}{p_{k,\ell}} - \frac{q_{k,\ell+1}}{p_{k,\ell+1}} - \frac{q_{k+1,\ell}}{p_{k+1,\ell}} + \frac{q_{k+1,\ell+1}}{p_{k+1,\ell+1}} \right| \left(\frac{P_{k,\ell}}{Q_{m,n}} \right) \\
&\quad + \sum_{\ell=0}^{n-1} \left| \frac{q_{m,\ell}}{p_{m,\ell}} - \frac{q_{m,\ell+1}}{p_{m,\ell+1}} \right| \left(\frac{P_{k,\ell}}{Q_{m,n}} \right) \\
&\quad + \sum_{k=0}^{m-1} \left| \frac{q_{k,n}}{p_{k,n}} - \frac{q_{k+1,n}}{p_{k+1,n}} \right| \left(\frac{P_{k,n}}{Q_{m,n}} \right) \\
&\quad + \left(\frac{q_{m,n}}{p_{m,n}} \right) \left(\frac{P_{m,n}}{Q_{m,n}} \right) \\
&\leq 4 \left(\frac{P_{m,n}}{Q_{m,n}} \right) \sum_{k,\ell=0}^{\infty,\infty} \left(\frac{q_{k,\ell}}{p_{k,\ell}} \right) \\
&\quad + 2 \left(\frac{P_{m,n}}{Q_{m,n}} \right) \sum_{k,\ell=0}^{\infty,\infty} \left(\frac{q_{k,\ell}}{p_{k,\ell}} \right) \\
&\quad + \left(\frac{P_{m,n}}{Q_{m,n}} \right) \sum_{k,\ell=0}^{\infty,\infty} \left(\frac{q_{k,\ell}}{p_{k,\ell}} \right) \\
&= 7 \left(\frac{P_{m,n}}{Q_{m,n}} \right) \sum_{k,\ell=0}^{\infty,\infty} \left(\frac{q_{k,\ell}}{p_{k,\ell}} \right) \\
&\leq 7M \sum_{k,\ell=0}^{\infty,\infty} \left(\frac{q_{k,\ell}}{p_{k,\ell}} \right), \quad m, n = 0, 1, 2, \ldots,
\end{aligned}$$

where $\frac{P_{m,n}}{Q_{m,n}} \leq M$, $m, n = 0, 1, 2, \ldots$, in view of (6.12). Thus,

$$\sup_{m,n\geq 0} \sum_{k,\ell=0}^{\infty,\infty} |c_{m,n,k,\ell}| < \infty.$$

We will now prove that for each fixed $\ell = 0, 1, 2, \ldots$,

$$\lim_{m+n\to\infty} \sum_{k=0}^{\infty} |c_{m,n,k,\ell}| = 0.$$

For such a fixed ℓ, we consider three cases, viz., $\ell < n$, $\ell = n$ and $\ell > n$.
Case 1. When $\ell < n$,

$$\begin{aligned}
\sum_{k=0}^{\infty} |c_{m,n,k,\ell}| &= \sum_{k=0}^{m-1} |c_{m,n,k,\ell}| + |c_{m,n,m,\ell}| \\
&= \sum_{k=0}^{m-1} \left| \frac{q_{k,\ell}}{p_{k,\ell}} - \frac{q_{k,\ell+1}}{p_{k,\ell+1}} - \frac{q_{k+1,\ell}}{p_{k+1,\ell}} + \frac{q_{k+1,\ell+1}}{p_{k+1,\ell+1}} \right| \left(\frac{P_{k,\ell}}{Q_{m,n}} \right) \\
&\quad + \left| \frac{q_{m,\ell}}{p_{m,\ell}} - \frac{q_{m,\ell+1}}{p_{m,\ell+1}} \right| \left(\frac{P_{m,\ell}}{Q_{m,n}} \right) \\
&\leq 6 \left(\frac{P_{m,n}}{Q_{m,n}} \right) \sum_{k,\ell=0}^{\infty,\infty} \left(\frac{q_{k,\ell}}{p_{k,\ell}} \right) \\
&\to 0, \ m+n \to \infty, \ \text{ in view of (6.11) and (6.12),}
\end{aligned}$$

proving that

$$\lim_{m+n\to\infty} \sum_{k=0}^{\infty} |c_{m,n,k,\ell}| = 0, \ \ell = 0, 1, 2, \ldots$$

in this case.
Case 2. The case $\ell = n$ can be proved similarly.
Case 3. If $\ell > n$, then $c_{m,n,k,\ell} = 0$, $m, n = 0, 1, 2, \ldots$ and so

$$\lim_{m+n\to\infty} \sum_{k=0}^{\infty} |c_{m,n,k,\ell}| = 0, \ \ell = 0, 1, 2, \ldots$$

in this case too. Consequently,

$$\lim_{m+n\to\infty} \sum_{k=0}^{\infty} |c_{m,n,k,\ell}| = 0, \ \ell = 0, 1, 2, \ldots$$

in all cases. Similarly, we can prove that

$$\lim_{m+n\to\infty} \sum_{\ell=0}^{\infty} |c_{m,n,k,\ell}| = 0, \; k = 0, 1, 2, \ldots.$$

Appealing to Theorem 5.3, it follows that the four-dimensional infinite matrix $(c_{m,n,k,\ell})$ is regular and so

$$(\overline{N}, p_{m,n}) \subseteq (\overline{N}, q_{m,n}),$$

completing the proof of the theorem.

Following the proof of Theorem 6.7, we can prove

Theorem 6.8 *If* $(\overline{N}, p_{m,n})$, $(\overline{N}, q_{m,n})$ *are two Weighted Mean methods such that*

$$q_{m,n} = O(p_{m,n}), \; m+n \to \infty, \tag{6.13}$$

i.e., there exists $H > 0$ *such that*

$$\frac{q_{m,n}}{p_{m,n}} \leq H, \; m, n = 0, 1, 2, \ldots$$

and

$$\sum_{m,n=0}^{\infty,\infty} \left(\frac{P_{m,n}}{Q_{m,n}}\right) < \infty, \tag{6.14}$$

then

$$(\overline{N}, p_{m,n}) \subseteq (\overline{N}, q_{m,n}).$$

We now prove another inclusion theorem.

Theorem 6.9 (Comparison theorem for a $(\overline{N}, p_{m,n})$ method and a regular matrix method) *Let* $(\overline{N}, p_{m,n})$ *be a Weighted Mean method and* $A = (a_{m,n,k,\ell})$ *be a regular four-dimensional infinite matrix method. If*

$$\sup_{m,n\geq 0} \sum_{k,\ell=0}^{\infty,\infty} |a_{m,n,k,\ell}| \frac{P_{k,\ell}}{p_{k,\ell}} < \infty; \tag{6.15}$$

$$\lim_{m+n\to\infty} \sum_{k=0}^{\infty} |a_{m,n,k,\ell}| \frac{P_{k,\ell}}{p_{k,\ell}} = 0, \; \ell = 0, 1, 2, \ldots; \tag{6.16}$$

and

$$\lim_{m+n\to\infty} \sum_{\ell=0}^{\infty} |a_{m,n,k,\ell}| \frac{P_{k,\ell}}{p_{k,\ell}} = 0, \; k = 0, 1, 2, \ldots, \tag{6.17}$$

then,

$$(\overline{N}, p_{m,n}) \subseteq A.$$

Proof Let $\{s_{m,n}\}$ be a double sequence and $\{t_{m,n}\}$, $\{\tau_{m,n}\}$ be its $(\overline{N}, p_{m,n})$ and A-transforms, respectively. Then,

$$t_{m,n} = \frac{\sum_{i,j=0}^{m,n} p_{i,j} s_{i,j}}{P_{m,n}},$$

$$\tau_{m,n} = \sum_{k,\ell=0}^{\infty,\infty} a_{m,n,k,\ell} s_{k,\ell}, \quad m, n = 0, 1, 2, \ldots.$$

Now,

$$s_{m,n} = \frac{P_{m,n}t_{m,n} - P_{m-1,n}t_{m-1,n} - P_{m,n-1}t_{m,n-1} + P_{m-1,n-1}t_{m-1,n-1}}{p_{m,n}}$$

$m, n = 0, 1, 2, \ldots$, where we suppose that

$$t_{-1,n} = t_{m,-1} = t_{-1,-1} = 0.$$

We now have

$$\begin{aligned}
\tau_{m,n} &= \sum_{k,\ell=0}^{\infty,\infty} a_{m,n,k,\ell} s_{k,\ell} \\
&= \sum_{k,\ell=0}^{\infty,\infty} \frac{a_{m,n,k,\ell}}{p_{k,\ell}} \left[P_{k,\ell}t_{k,\ell} - P_{k-1,\ell}t_{k-1,\ell} - P_{k,\ell-1}t_{k,\ell-1} + P_{k-1,\ell-1}t_{k-1,\ell-1}\right] \\
&= \sum_{k,\ell=0}^{\infty,\infty} \left[\frac{a_{m,n,k,\ell}}{p_{k,\ell}} - \frac{a_{m,n,k+1,\ell}}{p_{k+1,\ell}} - \frac{a_{m,n,k,\ell+1}}{p_{k,\ell+1}} + \frac{a_{m,n,k+1,\ell+1}}{p_{k+1,\ell+1}}\right] P_{k,\ell}t_{k,\ell}.
\end{aligned}$$

Let $\lim_{m+n\to\infty} t_{m,n} = s$. Since $\{t_{m,n}\}$ converges, it is bounded and so $|t_{m,n}| \leq M$, $m, n = 0, 1, 2, \ldots, M > 0$. Now,

$$\begin{aligned}
\sum_{k,\ell=0}^{\infty,\infty} |a_{m,n,k,\ell}| \frac{P_{k,\ell}}{p_{k,\ell}} |t_{k,\ell}| &\leq M \sum_{k,\ell=0}^{\infty,\infty} |a_{m,n,k,\ell}| \frac{P_{k,\ell}}{p_{k,\ell}} \\
&< \infty, \quad \text{in view of (6.15);}
\end{aligned}$$

$$\sum_{k,\ell=0}^{\infty,\infty} |a_{m,n,k+1,\ell}| \frac{P_{k,\ell}}{p_{k+1,\ell}} |t_{k,\ell}| \leq M \sum_{k,\ell=0}^{\infty,\infty} |a_{m,n,k+1,\ell}| \frac{P_{k+1,\ell}}{p_{k+1,\ell}},$$

$$\text{since } P_{k,\ell} \leq P_{k+1,\ell}$$

$$< \infty, \quad \text{in view of (6.15);}$$

Similarly,

$$\sum_{k,\ell=0}^{\infty,\infty} |a_{m,n,k,\ell+1}| \frac{P_{k,\ell}}{p_{k,\ell+1}} |t_{k,\ell}| < \infty$$

and

$$\sum_{k,\ell=0}^{\infty,\infty} |a_{m,n,k+1,\ell+1}| \frac{P_{k,\ell}}{p_{k+1,\ell+1}} |t_{k,\ell}| < \infty,$$

in view of (6.15). We now write

$$\tau_{m,n} = \sum_{k,\ell=0}^{\infty,\infty} b_{m,n,k,\ell} t_{k,\ell}, \quad m, n = 0, 1, 2, \ldots,$$

where

$$b_{m,n,k,\ell} = \left[\frac{a_{m,n,k,\ell}}{p_{k,\ell}} - \frac{a_{m,n,k+1,\ell}}{p_{k+1,\ell}} - \frac{a_{m,n,k,\ell+1}}{p_{k,\ell+1}} + \frac{a_{m,n,k+1,\ell+1}}{p_{k+1,\ell+1}} \right] P_{k,\ell},$$

$m, n, k, \ell = 0, 1, 2, \ldots$. Now,

$$\sum_{k,\ell=0}^{\infty,\infty} |b_{m,n,k,\ell}| \leq 4 \sup_{m,n \geq 0} \sum_{k,\ell=0}^{\infty,\infty} |a_{m,n,k,\ell}| \frac{P_{k,\ell}}{p_{k,\ell}}$$

$$< \infty,$$

$m, n = 0, 1, 2, \ldots$, in view of (6.15), so that

$$\sup_{m,n \geq 0} \sum_{k,\ell=0}^{\infty,\infty} |b_{m,n,k,\ell}| < \infty.$$

Since A is regular,

$$\lim_{m+n \to \infty} a_{m,n,k,\ell} = 0, \quad k, \ell = 0, 1, 2, \ldots$$

in view of Theorem 5.3, from which it follows that

$$\lim_{m+n \to \infty} b_{m,n,k,\ell} = 0, \quad k, \ell = 0, 1, 2, \ldots.$$

Let $s_{m,n} = 1, m, n = 0, 1, 2, \ldots$. Then, $t_{m,n} = 1, m, n = 0, 1, 2, \ldots$. Now,

$$\sum_{k,\ell=0}^{\infty,\infty} b_{m,n,k,\ell} = \sum_{k,\ell=0}^{\infty,\infty} a_{m,n,k,\ell}, \quad m, n = 0, 1, 2, \ldots,$$

so that

$$\begin{aligned} \lim_{m+n\to\infty} \sum_{k,\ell=0}^{\infty,\infty} b_{m,n,k,\ell} &= \lim_{m+n\to\infty} \sum_{k,\ell=0}^{\infty,\infty} a_{m,n,k,\ell} \\ &= 1, \end{aligned}$$

since A is regular, again appealing to Theorem 5.3. Also, for every fixed $\ell = 0, 1, 2, \ldots$,

$$\begin{aligned} &\lim_{m+n\to\infty} \sum_{k=0}^{\infty} |b_{m,n,k,\ell}| \\ &= \lim_{m+n\to\infty} \sum_{k=0}^{\infty} \left| \frac{a_{m,n,k,\ell}}{p_{k,\ell}} - \frac{a_{m,n,k+1,\ell}}{p_{k+1,\ell}} - \frac{a_{m,n,k,\ell+1}}{p_{k,\ell+1}} + \frac{a_{m,n,k+1,\ell+1}}{p_{k+1,\ell+1}} \right| P_{k,\ell} \\ &\leq \lim_{m+n\to\infty} \left[\sum_{k=0}^{\infty} |a_{m,n,k,\ell}| \frac{P_{k,\ell}}{p_{k,\ell}} + \sum_{k=0}^{\infty} |a_{m,n,k+1,\ell}| \frac{P_{k,\ell}}{p_{k+1,\ell}} \right. \\ &\left. + \sum_{k=0}^{\infty} |a_{m,n,k,\ell+1}| \frac{P_{k,\ell}}{p_{k,\ell+1}} + \sum_{k=0}^{\infty} |a_{m,n,k+1,\ell+1}| \frac{P_{k,\ell}}{p_{k+1,\ell+1}} \right] \\ &\leq 4 \lim_{m+n\to\infty} \sum_{k=0}^{\infty} |a_{m,n,k,\ell}| \frac{P_{k,\ell}}{p_{k,\ell}} \\ &= 0, \quad \text{using (6.16)}. \end{aligned}$$

Thus,

$$\lim_{m+n\to\infty} \sum_{k=0}^{\infty} |b_{m,n,k,\ell}| = 0, \quad \ell = 0, 1, 2, \ldots.$$

Similarly, we can prove that

$$\lim_{m+n\to\infty} \sum_{\ell=0}^{\infty} |b_{m,n,k,\ell}| = 0, \quad k = 0, 1, 2, \ldots,$$

using (6.17). Consequently, the four-dimensional infinite matrix $(b_{m,n,k,\ell})$ is regular, again using Theorem 5.3. So $\lim_{k+\ell\to\infty} t_{k,\ell} = s$ implies that $\lim_{m+n\to\infty} \tau_{m,n} = s$. In other words,

$$(\overline{N}, p_{m,n}) \subseteq A.$$

The proof of the theorem is now complete. □

Corollary 6.1 *If we choose a regular Weighted Mean method $(\overline{N}, q_{m,n})$ for A, we get the corresponding sufficient conditions, using (6.15)–(6.17), for*

$$(\overline{N}, p_{m,n}) \subseteq (\overline{N}, q_{m,n}).$$

6.3 $(M, \lambda_{m,n})$ or Natarajan Method for Double Sequences

In this section, we introduce the $(M, \lambda_{m,n})$ method or the Natarajan method for double sequences and extend some of the results of the (M, λ_n) method (or the Natarajan method) for simple sequences presented in Chap. 5 (for details, see [5]).

Definition 6.8 Let $\{\lambda_{m,n}\}$ be a double sequence such that $\sum_{m,n=0}^{\infty,\infty} |\lambda_{m,n}| < \infty$. The $(M, \lambda_{m,n})$ method is defined by the four-dimensional infinite matrix $(a_{m,n,k,\ell})$, where

$$a_{m,n,k,\ell} = \begin{cases} \lambda_{m-k,n-\ell}, & k \leq m, \ell \leq n; \\ 0, & \text{otherwise.} \end{cases}$$

Definition 6.9 The methods $(M, \lambda_{m,n})$, $(M, \mu_{m,n})$ are said to be consistent, if

$$s_{k,\ell} \to \sigma(M, \lambda_{m,n}) \text{ and } s_{k,\ell} \to \sigma'(M, \mu_{m,n}) \text{ imply that } \sigma = \sigma'.$$

Definition 6.10 We say that $(M, \lambda_{m,n})$ is included in $(M, \mu_{m,n})$ (or $(M, \mu_{m,n})$ includes $(M, \lambda_{m,n})$), written as

$$(M, \lambda_{m,n}) \subseteq (M, \mu_{m,n}) \text{ (or } (M, \mu_{m,n}) \supseteq (M, \lambda_{m,n}))$$

if

$$s_{k,\ell} \to \sigma(M, \lambda_{m,n}) \text{ implies that } s_{k,\ell} \to \sigma(M, \mu_{m,n}) \text{ too.}$$

The methods $(M, \lambda_{m,n})$, $(M, \mu_{m,n})$ are said to be equivalent if

$$(M, \lambda_{m,n}) \subseteq (M, \mu_{m,n}) \text{ and vice versa.}$$

It is easy to prove the following result.

Theorem 6.10 *The method $(M, \lambda_{m,n})$ is regular if and only if*

$$\sum_{m,n=0}^{\infty,\infty} \lambda_{m,n} = 1.$$

In the sequel, let $(M, \lambda_{m,n})$, $(M, \mu_{m,n})$ be regular methods such that each row and each column of the two-dimensional infinite matrices $(\lambda_{m,n})$, $(\mu_{m,n})$ is a regular Natarajan method for simple sequences.

Theorem 6.11 *Any two such regular Natarajan methods are consistent.*

Proof Let $(M, \lambda_{m,n})$, $(M, \mu_{m,n})$ be two regular methods such that each row and each column of the two-dimensional infinite matrices $(\lambda_{m,n})$, $(\mu_{m,n})$ is a regular Natarajan method for simple sequences. We now define a third method $(M, \gamma_{m,n})$ by the equation

$$\gamma_{m,n} = \sum_{i,j=0}^{m,n} \lambda_{i,j}\mu_{m-i,n-j}, \quad m, n = 0, 1, 2, \ldots .$$

Now, for $s = \{s_{m,n}\}$, we get

$$(M, \gamma_{m,n})(s) = \sum_{i,j=0}^{\infty,\infty} u_{m,n,i,j}(M, \mu_{i,j})(s),$$

where

$$u_{m,n,i,j} = \begin{cases} \lambda_{m-i,n-j}Q_{i,j}, & \text{if } i \leq m, \ j \leq n; \\ 0, & \text{otherwise,} \end{cases}$$

$Q_{i,j} = \sum_{k,\ell=0}^{i,j} \mu_{k,\ell}$, $i, j = 0, 1, 2, \ldots$. Using Theorem 6.8, we can verify that the four-dimensional infinite matrix $(u_{m,n,i,j})$ is regular. Thus,

$$s_{k,\ell} \to \sigma'(M, \mu_{m,n}) \text{ implies that } s_{k,\ell} \to \sigma'(M, \gamma_{m,n}).$$

Similarly, we can prove that

$$s_{k,\ell} \to \sigma(M, \lambda_{m,n}) \text{ implies that } s_{k,\ell} \to \sigma(M, \gamma_{m,n}).$$

Consequently, $\sigma = \sigma'$ and so the methods $(M, \lambda_{m,n})$ and $(M, \mu_{m,n})$ are consistent, completing the proof of the theorem. □

Let

$$\lambda(x, y) = \sum_{m,n=0}^{\infty,\infty} \lambda_{m,n}x^m y^n, \quad \mu(x, y) = \sum_{m,n=0}^{\infty,\infty} \mu_{m,n}x^m y^n.$$

It is clear that these series converge for $|x|, |y| < 1$. Let

$$k(x, y) = \sum_{m,n=0}^{\infty,\infty} k_{m,n} x^m y^n = \frac{\mu(x, y)}{\lambda(x, y)};$$

$$h(x, y) = \sum_{m,n=0}^{\infty,\infty} h_{m,n} x^m y^n = \frac{\lambda(x, y)}{\mu(x, y)}.$$

We note that

$$\mu_{m,n} = \sum_{i,j=0}^{m,n} k_{i,j} \lambda_{m-i,n-j};$$

$$\lambda_{m,n} = \sum_{i,j=0}^{m,n} h_{i,j} \mu_{m-i,n-j},$$

$m, n = 0, 1, 2, \ldots$.

We now have

Theorem 6.12 *If $(M, \lambda_{m,n})$, $(M, \mu_{m,n})$ are regular, then*

$$(M, \lambda_{m,n}) \subseteq (M, \mu_{m,n})$$

if and only if

$$\sum_{m,n=0}^{\infty,\infty} |k_{m,n}| < \infty \text{ and } \sum_{m,n=0}^{\infty,\infty} k_{m,n} = 1.$$

Proof Let

$$s(x, y) = \sum_{m,n=0}^{\infty,\infty} s_{m,n} x^m y^n.$$

Then,

$$\sum_{m,n=0}^{\infty,\infty} (M, \mu_{m,n})(s) x^m y^n = \sum_{m,n=0}^{\infty,\infty} \left(\sum_{i,j=0}^{m,n} \mu_{m-i,n-j} s_{i,j} \right) x^m y^n$$
$$= \mu(x, y) s(x, y).$$

Similarly,

$$\sum_{m,n=0}^{\infty,\infty} (M, \lambda_{m,n})(s) x^m y^n = \lambda(x, y) s(x, y).$$

Thus,

$$\sum_{m,n=0}^{\infty,\infty} (M, \mu_{m,n})(s)x^m y^n = k(x, y) \sum_{m,n=0}^{\infty,\infty} (M, \lambda_{m,n})(s)x^m y^n$$
$$= \left(\sum_{m,n=0}^{\infty,\infty} k_{m,n}x^m y^n\right)\left(\sum_{m,n=0}^{\infty,\infty} (M, \lambda_{m,n})(s)x^m y^n\right),$$

which implies that

$$(M, \mu_{m,n})(s) = \sum_{i,j=0}^{m,n} k_{m-i,n-j}(M, \lambda_{i,j})(s)$$
$$= \sum_{i,j=0}^{\infty,\infty} c_{m,n,i,j}(M, \lambda_{i,j})(s),$$

where

$$c_{m,n,i,j} = \begin{cases} k_{m-i,n-j}, & \text{if } i \leq m,\ j \leq n; \\ 0, & \text{otherwise.} \end{cases}$$

If $(M, \lambda_{m,n}) \subseteq (M, \mu_{m,n})$, then $(c_{m,n,i,j})$ is regular. In view of Theorem 5.3,

$$\lim_{m+n\to\infty} \sum_{i,j=0}^{\infty,\infty} c_{m,n,i,j} = 1,$$
$$i.e., \lim_{m+n\to\infty} \sum_{i,j=0}^{m,n} k_{m-i,n-j} = 1,$$
$$i.e., \lim_{m+n\to\infty} \sum_{i,j=0}^{m,n} k_{i,j} = 1,$$
$$i.e., \sum_{m,n=0}^{\infty,\infty} k_{m,n} = 1.$$

Again, by Theorem 5.3, there is $H > 0$ such that

$$\sum_{i,j=0}^{\infty,\infty} |c_{m,n,i,j}| \leq H, \quad m, n = 0, 1, 2, \ldots,$$
$$i.e., \sum_{i,j=0}^{m,n} |k_{m-i,n-j}| \leq H, \quad m, n = 0, 1, 2, \ldots,$$

$$i.e., \sum_{i,j=0}^{m,n} |k_{i,j}| \leq H, \quad m, n = 0, 1, 2, \ldots,$$

from which it follows that

$$\sum_{m,n=0}^{\infty,\infty} |k_{m,n}| < \infty.$$

Conversely, if $\sum_{m,n=0}^{\infty,\infty} |k_{m,n}| < \infty$ and $\sum_{m,n=0}^{\infty,\infty} k_{m,n} = 1$, it is easy to verify that $(c_{m,n,i,j})$ is regular and consequently,

$$(M, \lambda_{m,n}) \subseteq (M, \mu_{m,n}),$$

completing the proof of the theorem. □

As a consequence of Theorem 6.12, we have

Theorem 6.13 *The regular methods $(M, \lambda_{m,n})$, $(M, \mu_{m,n})$ are equivalent if and only if*

$$\sum_{m,n=0}^{\infty,\infty} |k_{m,n}| < \infty, \sum_{m,n=0}^{\infty,\infty} k_{m,n} = 1$$

and

$$\sum_{m,n=0}^{\infty,\infty} |h_{m,n}| < \infty, \sum_{m,n=0}^{\infty,\infty} h_{m,n} = 1.$$

Analogous to Theorem 3.14, we have the following result in the context of double sequences and double series:

Theorem 6.14 *If* $\lim_{m+n\to\infty} a_{m,n} = 0$ *and* $\sum_{m,n=0}^{\infty,\infty} |b_{m,n}| < \infty$, *then,*

$$\lim_{m+n\to\infty} \sum_{k,\ell=0}^{m,n} a_{m-k,n-\ell} b_{k,\ell} = 0.$$

Proof Since $\{a_{m,n}\}, \{b_{m,n}\}$ are convergent, they are bounded and so, there exits $M > 0$ such that

$$|a_{m,n}| \leq M, \ |b_{m,n}| \leq M, \ m, n = 0, 1, 2, \ldots.$$

Since $\sum_{m,n=0}^{\infty,\infty} |b_{m,n}| < \infty$, given $\epsilon > 0$, there exist positive integers M_1, N_1 such that

$$\sum_{m>M_1,n>N_1}^{\infty,\infty} |b_{m,n}| < \frac{\epsilon}{4M}. \tag{6.18}$$

Since, for fixed $k, \ell = 0, 1, 2, \ldots,$

$$\lim_{m+n\to\infty} a_{m-k,n-\ell} = 0,$$

we can choose positive integers $M_2 > M_1$, $N_2 > N_1$ such that for $m > M_2$, $n > N_2$, we have

$$\sup_{\substack{0\le k\le M_1\\ 0\le \ell\le N_1}} |a_{m-k,n-\ell}| < \frac{\epsilon}{4M}; \tag{6.19}$$

$$\sup_{\substack{0\le k\le M_1\\ N_1+1\le \ell\le n}} |a_{m-k,n-\ell}| < \frac{\epsilon}{4M}; \tag{6.20}$$

and

$$\sup_{\substack{M_1+1\le k\le m\\ 0\le \ell\le N_1}} |a_{m-k,n-\ell}| < \frac{\epsilon}{4M}. \tag{6.21}$$

Then, for $m > M_2$, $n > N_2$,

$$\begin{aligned}
&\left|\sum_{k,\ell=0}^{m,n} a_{m-k,n-\ell} b_{k,\ell}\right| \\
&= \left|\sum_{\substack{0\le k\le M_1\\ 0\le \ell\le N_1}} a_{m-k,n-\ell} b_{k,\ell} + \sum_{\substack{0\le k\le M_1\\ \ell> N_1}} a_{m-k,n-\ell} b_{k,\ell}\right. \\
&\quad \left. + \sum_{\substack{k>M_1\\ 0\le \ell\le N_1}} a_{m-k,n-\ell} b_{k,\ell} + \sum_{\substack{k>M_1\\ \ell> N_1}} a_{m-k,n-\ell} b_{k,\ell}\right| \\
&= \sum_{\substack{0\le k\le M_1\\ 0\le \ell\le N_1}} |a_{m-k,n-\ell}||b_{k,\ell}| + \sum_{\substack{0\le k\le M_1\\ \ell> N_1}} |a_{m-k,n-\ell}||b_{k,\ell}| \\
&\quad + \sum_{\substack{k>M_1\\ 0\le \ell\le N_1}} |a_{m-k,n-\ell}||b_{k,\ell}| + \sum_{\substack{k>M_1\\ \ell> N_1}} |a_{m-k,n-\ell}||b_{k,\ell}| \\
&< M\frac{\epsilon}{4M} + M\frac{\epsilon}{4M} + M\frac{\epsilon}{4M} + M\frac{\epsilon}{4M} \\
&= \epsilon, \quad \text{using (6.18)–(6.21).}
\end{aligned}$$

It now follows that

$$\lim_{m+n\to\infty} \sum_{k,\ell=0}^{m,n} a_{m-k,n-\ell} b_{k,\ell} = 0,$$

completing the proof of the theorem. □

We now have the following results on the Cauchy multiplication of $(M, \lambda_{m,n})$-summable double sequences and double series (see [6]).

Theorem 6.15 *If* $\sum_{m,n=0}^{\infty,\infty} |a_{m,n}| < \infty$ *and* $\{b_{m,n}\}$ *is* $(M, \lambda_{m,n})$*-summable to* B*, then* $\{c_{m,n}\}$ *is* $(M, \lambda_{m,n})$*-summable to* AB*, where*

$$c_{m,n} = \sum_{k,\ell=0}^{m,n} a_{m-k,n-\ell} b_{k,\ell}, \quad m, n = 0, 1, 2, \ldots$$

and $\sum_{m,n=0}^{\infty,\infty} a_{m,n} = A$.

Proof Let $\{t_{m,n}\}$, $\{u_{m,n}\}$ be the $(M, \lambda_{m,n})$-transforms of $\{b_{m,n}\}$, $\{c_{m,n}\}$, respectively,

$$i.e., \ t_{m,n} = \sum_{k,\ell=0}^{m,n} \lambda_{m-k,n-\ell} b_{k,\ell},$$

$$u_{m,n} = \sum_{k,\ell=0}^{m,n} \lambda_{m-k,n-\ell} c_{k,\ell}, \quad m, n = 0, 1, 2, \ldots.$$

We can work to see that

$$u_{m,n} = \sum_{k,\ell=0}^{m,n} a_{m-k,n-\ell}(t_{k,\ell} - B) + B\left(\sum_{k,\ell=0}^{m,n} a_{k,\ell}\right), \quad m, n = 0, 1, 2, \ldots,$$

where $\lim_{k+\ell\to\infty} t_{k,\ell} = B$. Since $\sum_{m,n=0}^{\infty,\infty} |a_{m,n}| < \infty$ and $\lim_{m+n\to\infty} t_{m,n} = B$, using Theorem 6.14, it follows that

$$\lim_{m+n\to\infty} \left[\sum_{k,\ell=0}^{m,n} a_{m-k,n-\ell}(t_{k,\ell} - B)\right] = 0,$$

so that

$$\lim_{m+n\to\infty} u_{m,n} = B\left(\sum_{m,n=0}^{\infty,\infty} a_{m,n}\right) = AB,$$

completing the proof of the theorem. □

It is easy to prove the following result on similar lines:

Theorem 6.16 *If* $\sum_{m,n=0}^{\infty,\infty} |a_{m,n}| < \infty$ *with* $\sum_{m,n=0}^{\infty,\infty} a_{m,n} = A$ *and* $\sum_{m,n=0}^{\infty,\infty} b_{m,n}$ *is* $(M, \lambda_{m,n})$-*summable to* B, *then* $\sum_{m,n=0}^{\infty,\infty} c_{m,n}$ *is* $(M, \lambda_{m,n})$-*summable to* AB, *where*

$$c_{m,n} = \sum_{k,\ell=0}^{m,n} a_{m-k,n-\ell} b_{k,\ell}, \quad m, n = 0, 1, 2, \ldots.$$

As in the case of the Natarajan method (M, λ_n) for simple sequences, we can prove the following result, using Theorem 6.14.

Theorem 6.17 *Let* $(M, \lambda_{m,n})$, $(M, \mu_{m,n})$ *be regular methods. Then* $(M, \lambda_{m,n})$ $(M, \mu_{m,n})$ *is also regular, where we define, for* $x = \{x_{m,n}\}$,

$$((M, \lambda_{m,n})(M, \mu_{m,n}))(x) = (M, \lambda_{m,n})((M, \mu_{m,n})(x)).$$

We can prove the following results too.

Theorem 6.18 *For given regular methods* $(M, \lambda_{m,n})$, $(M, \mu_{m,n})$, *and* $(M, t_{m,n})$,

$$(M, \lambda_{m,n}) \subseteq (M, \mu_{m,n})$$

if and only if

$$(M, t_{m,n})(M, \lambda_{m,n}) \subseteq (M, t_{m,n})(M, \mu_{m,n}).$$

In view of Theorem 6.12, Theorem 6.18 can be reformulated as under:

Theorem 6.19 *Given the regular methods* $(M, \lambda_{m,n})$, $(M, \mu_{m,n})$ *and* $(M, t_{m,n})$, *the following statements are equivalent:*

(i) $(M, \lambda_{m,n}) \subseteq (M, \mu_{m,n})$;
(ii) $(M, t_{m,n})(M, \lambda_{m,n}) \subseteq (M, t_{m,n})(M, \mu_{m,n})$;

and

(iii) $\sum_{m,n=0}^{\infty,\infty} |k_{m,n}| < \infty$ *and* $\sum_{m,n=0}^{\infty,\infty} k_{m,n} = 1$,

where

$$\frac{\mu(x)}{\lambda(x)} = k(x) = \sum_{m,n=0}^{\infty,\infty} k_{m,n} x^m y^n,$$

$$\lambda(x) = \sum_{m,n=0}^{\infty,\infty} \lambda_{m,n} x^m y^n,$$

$$\mu(x) = \sum_{m,n=0}^{\infty,\infty} \mu_{m,n} x^m y^n.$$

In this context, we remark that we can extend many more results for the Nörlund, the Weighted Mean, and the Natarajan methods for simple sequences to the corresponding summability methods for double sequences. It is left to the reader to try and solve such problems.

While concluding the present book, the author wishes to point out that the summability of trigonometric and Walsh–Fourier series has been studied intensively in the literature in the past 20–30 years (see, for instance, [7–12]). The readers, who are interested in summability theory, can acquaint themselves in this part of the literature too for further studies.

References

1. Natarajan, P.N.: Nörlund means for double sequences and double series (communicated for publication)
2. Moore, C.N.: On relationships between Nörlund means for double series. Proc. Amer. Math. Soc. **5**, 957–963 (1954)
3. Natarajan, P.N.: Weighted means for double sequences and double series. Indian J. Math. **58**, 31–42 (2016)
4. Natarajan, P.N.: An inclusion theorem for weighted mean methods for double sequences. Indian J. Math. (Supplement) Proceedings Sixth Dr. George Bachman Memorial Conference, **57**, 39–42 (2015)
5. Natarajan, P.N.: Natarajan method of summability for double sequences and double series. An. Stiint. Univ. Al. I. Cuza Iasi Math. (N.S.) Tomul LXII, **2**, 547–552 (2016)
6. Natarajan, P.N.: New properties of the Natarajan method of summability for double sequences. Int. J. Phys. Math. Sci. **6**(3), 28–33 (2016)
7. Butzer, P.L., Nessel, R.J.: Fourier Analysis and Applications. Birkhauser, Basel (1971)
8. Schipp, F., Wade, W.R., Simon, P., Pal, J.: Walsh Series: An Introduction to Dyadic Harmonic Analysis. Adam Hilger, Bristol (1990)
9. Shawyer, B., Watson, B.: Borel's Methods of Summability. Oxford (1994)
10. Weisz, F.: Summability of Multidimensional Fourier series and Hardy spaces. Mathematics and its Applications. Kluwer Academic Publishers, Dordrecht (2002)
11. Zygmund, A.: Trigonometric Series, 3rd edn. Cambridge, London (2002)
12. Grafakos, L.: Classical and Modern Fourier Analysis. Pearson Education, New Jersey (2004)

Index

P.N. Natarajan, *Classical Summability Theory*, DOI 10.1007/978-981-10-4205-8